PERIODIC TABLE
OF THE **ELEMENTS**

Alkali &
Alkaline
Earth Metals

PERIODIC TABLE
OF THE ELEMENTS

Alkali &
Alkaline
Earth Metals

Monica Halka, Ph.D., and
Brian Nordstrom, Ed.D.

An imprint of Infobase Publishing

Facts On File, Inc.
An imprint of Infobase Publishing
132 West 31st Street
New York NY 10001

Library of Congress Cataloging-in-Publication Data
Halka, Monica.
  Alkali and alkaline earth metals / Monica Halka and Brian Nordstrom.
      p. cm. — (Periodic table of the elements)
  Includes bibliographical references and index.
  ISBN 978-0-8160-7369-6
  1. Alkalies. 2. Alkaline earth metals. 3. Periodic law. I. Nordstrom, Brian. II. Title.
  QD172.A4H35 2010
  546'.38—dc22   2009035152

Facts On File books are available at special discounts when purchased in bulk quantities for businesses, associations, institutions, or sales promotions. Please call our Special Sales Department in New York at (212) 967-8800 or (800) 322-8755.

You can find Facts On File on the World Wide Web at http://www.factsonfile.com

Excerpts included herewith have been reprinted by permission of the copyright holders; the author has made every effort to contact copyright holders. The publishers will be glad to rectify, in future editions, any errors or omissions brought to their notice.

Text design by Erik Lindstrom
Composition by Hermitage Publishing Services
Illustrations by Dale Williams
Photo research by Tobi Zausner, Ph.D.
Cover printed by Yurchak Printing, Landisville, Pa.
Book printed and bound by Yurchak Printing, Landisville, Pa.
Printed in the United States of America

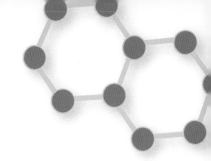

Contents

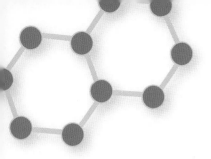

Preface

Speculations about the nature of matter date back to ancient Greek philosophers like Thales, who lived in the sixth century B.C.E., and Democritus, who lived in the fifth century B.C.E., and to whom we credit the first theory of *atoms*. It has taken two and a half millennia for natural philosophers and, more recently, for chemists and physicists to arrive at a modern understanding of the nature of *elements* and *compounds*. By the 19th century, chemists such as John Dalton of England had learned to define elements as pure substances that contain only one kind of atom. It took scientists like the British physicists Joseph John Thomson and Ernest Rutherford in the early years of the 20th century, however, to demonstrate what atoms are—entities composed of even smaller and more elementary particles called *protons, neutrons,* and *electrons.* These particles give atoms their properties and, in turn, give elements their physical and chemical properties.

After Dalton, there were several attempts throughout Western Europe to organize the known elements into a conceptual framework that would account for the similar properties that related groups of elements exhibit and for trends in properties that correlate with increases in atomic weights. The most successful *periodic table* of the elements was designed in 1869 by a Russian chemist, Dmitri Mendeleev. Mendeleev's method of organizing the elements into columns grouping elements with similar chemical and physical properties proved to be so practical that his table is still essentially the only one in use today.

While there are many excellent works written about the periodic table (which are listed in the section on further resources), recent scientific investigation has uncovered much that was previously unknown about nearly every element. The Periodic Table of the Elements, a six-volume set, is intended not only to explain how the elements were discovered and what their most prominent chemical and physical properties are, but also to inform the reader of new discoveries and uses in fields ranging from astrophysics to material science. Students, teachers, and the general public seldom have the opportunity to keep abreast of these new developments, as journal articles for the nonspecialist are hard to find. This work attempts to communicate new scientific findings simply and clearly, in language accessible to readers with little or no formal background in chemistry or physics. It should, however, also appeal to scientists who wish to update their understanding of the natural elements.

Each volume highlights a group of related elements as they appear in the periodic table. For each element, the set provides information regarding:

- the discovery and naming of the element, including its role in history, and some (though not all) of the important scientists involved;
- the basics of the element, including such properties as its atomic number, atomic mass, electronic configuration, melting and boiling temperatures, abundances (when known), and important isotopes;
- the chemistry of the element;
- new developments and dilemmas regarding current understanding; and
- past, present, and possible future uses of the element in science and technology.

Some topics, while important to many elements, do not apply to all. Though nearly all elements are known to have originated in stars or stellar explosions, little information is available for some. Some others that have been synthesized by scientists on Earth have not been observed in stellar spectra. If significant astrophysical nucleosynthesis research exists, it is presented as a separate section. The similar situation applies for geophysical research.

Special topic sections describe applications for two or more closely associated elements. Sidebars mainly refer to new developments of special interest. Further resources for the reader appear at the end of the book, with specific listings pertaining to each chapter, as well as a listing of some more general resources.

Acknowledgments

First and foremost, I thank my parents, who convinced me that I was capable of achieving any goal. In graduate school, my thesis adviser, Dr. Howard Bryant, influenced my way of thinking about science more than anyone else. Howard taught me that learning requires having the humility to doubt your understanding and that it is important for a physicist to be able to explain her work to anyone. I have always admired the ability to communicate scientific ideas to nonscientists and wish to express my appreciation for conversations with National Public Radio science correspondent Joe Palca, whose clarity of style I attempt to emulate in this work. I also thank my coworkers at Georgia Tech, Dr. Greg Nobles and Ms. Nicole Leonard, for their patience and humor as I struggled with deadlines.

—*Monica Halka*

In 1967, I entered the University of California at Berkeley. Several professors, including John Phillips, George Trilling, Robert Brown, Samuel Markowitz, and A. Starker Leopold, made significant and lasting impressions. I owe an especial debt of gratitude to Harold Johnston, who was my graduate research adviser in the field of atmospheric chemistry. I have known personally many of the scientists mentioned in the Periodic Table of the Elements set: For example, I studied under Neil Bartlett, Kenneth Street, Jr., and physics Nobel laureate Emilio Segrè.

I especially cherish having known chemistry Nobel laureate Glenn Seaborg. I also acknowledge my past and present colleagues at California State University; Northern Arizona University; and Embry-Riddle Aeronautical University, Prescott, Arizona, without whom my career in education would not have been as enjoyable.

—*Brian Nordstrom*

Both authors thank Jodie Rhodes and Frank Darmstadt for their encouragement, patience, and understanding.

Introduction

Materials that are good conductors of electricity are generally considered metals. One important use of metals, in fact, is the capability to be used in electrical circuitry. All of the metallic elements on Earth exist in its crust, mantle, or core. In addition, many of the metals that comprise the subject of this book are found as dissolved salts in seawater.

While scientists categorize the chemical elements as metals, nonmetals, and metalloids largely based on the elements' abilities to conduct electricity at normal temperatures and pressures, there are other distinctions taken into account when classifying the elements in the periodic table. The alkali metals, for example, are metals, but have such special properties that they are given their own classification. The same is true for the alkaline earths. Both families of elements appear in the two columns on the far left side of the periodic table. (See the following table, which shows the relative positions of the alkali metals and alkaline earths compared with the metals in columns IIIB, IVB, and VB in the periodic table on page 124.)

Alkali and Alkaline Earth Metals presents the current scientific understanding of the physics, chemistry, geology, and biology of these two families of elements, including how they are synthesized in the universe, when and how they were discovered, and where they are found on Earth. The book also details how humans use alkalis and alkaline earths and the resulting benefits and challenges to society, health, and the environment.

THE ALKALI METALS AND ALKALINE EARTH METALS

H				
Li	Be			
Na	Mg			
K	Ca	Sc	Ti	V
Rb	Sr	Y	Zr	Nb
Cs	Ba	La	Hf	Ta
Fr	Ra	Ac	Th	Pa

Note: Alkali metals are in italics. Alkaline earth metals are in bold type.

The first chapter discusses lithium, the lightest metal. Lithium is much in the news because of its current and anticipated future use in lightweight batteries.

Chapters 2 and 3 discuss two elements that are essential to human health—sodium and potassium, respectively. Sodium and potassium salts are the two most important electrolytes in the human body, and are responsible for ion and nervous-transport processes upon which life depends.

Chapter 4 examines the heavier alkali metals—rubidium, cesium, and francium. Francium is a radioactive, rare element; its longest-lived isotope has a half-life of only 22 minutes. The relative abundances of rubidium and cesium are much less than the abundances of lithium, sodium, or potassium, yet rubidium and cesium find important applications in atomic clocks and laser technology.

The subject of chapter 5 is beryllium, the lightest metal that can be used in structural materials. Beryllium is important in the aerospace industry, where its light weight contributes to lighter weight aircraft and spacecraft structures. Beryllium is also important in the nuclear power and weapons industries.

Chapters 6 and 7 investigate two more elements that are essential to human health—magnesium and calcium, respectively. Magnesium and calcium are found in several common minerals such as dolomite, calcite, limestone, and gypsum, and they are obtained from the evaporation of seawater. Calcium is an essential component of bones and teeth. Both elements are components of the electrolytes required by the body to maintain normal metabolic processes.

Chapter 8 discusses the heavier alkaline earth elements strontium and barium. Neither element plays a role in human health. These two elements occur in much smaller relative abundances than magnesium or calcium, and therefore find fewer, but nevertheless important, applications.

Chapter 9 covers radium, which exists only in radioactive forms. Radium has a fascinating history, from its discovery by Marie Curie to its applications in nuclear medicine.

Chapter 10 explains the chemistry and physics that underlie the basic properties of the alkali and the alkaline earth metals. In addition, it presents possible future developments that involve these two families of elements.

As an important introductory tool, the reader should note the following properties of metals in general:

1. The atoms of metals tend to be larger than those of non-metals. Several of the properties of metals result from their atomic sizes.

2. Metals exhibit high electrical conductivities. High electrical conductivity is the most important property that distinguishes metals from nonmetals.

3. Metals have low electronegativities; in fact, they are electropositive. This means that the atoms of metals have a strong tendency to lose electrons to form positively charged ions, a tendency that is responsible for metals' electrical conductivities.

4. Metals have low electron affinities. This means that gaining additional electrons is energetically unfavorable. Metal atoms would much rather give up one or more electrons than gain electrons.

5. Under normal conditions of temperature and pressure, with the exception of mercury, all metals are solids at room temperature. In contrast, many nonmetals are gases, one is a liquid, and only a few are solids. The fact that so many metals

exist as solids means that metals generally have relatively high melting and boiling points under normal atmospheric conditions.

6. In their solid state, metals tend to be malleable and ductile. They can be shaped or hammered into sheets, and they can be drawn into wires.

7. Metals tend to be shiny, or lustrous.

Alkali metals and alkaline earths have many similar properties. The following is a list of the general chemical and physical properties of these two families:

1. None of these elements can be found in nature as the pure elements; they all exist as compounds.

2. Alkali and alkaline metals are the most reactive metals in the periodic table. All of these elements react readily with water; rubidium and cesium do so explosively.

3. Alkali metals are very soft; the heavier ones can be cut with a butter knife. The alkaline earths tend to be harder metals.

4. For metals, these elements have relatively low melting points.

5. The densities of lithium and sodium are low enough that they float on water. The other elements in these two families are denser than water.

6. All of these elements can be identified using flame tests; when heated, each alkali or alkaline earth ion glows in the visible part of the spectrum, emitting light that is violet, blue, green, yellow, orange, or red, depending on the element.

7. Elements in both families only form simple positive ions— +1 ions in the case of alkalis and +2 in the case of alkaline earths. These ions easily form compounds with nonmetallic elements.

8. The +1 and +2 ions also combine with negatively charged hydrogen ions to form hydrides.

9. These elements never form polyatomic ions, nor do they form ions with charges higher than +1 or +2.

10. Alkalis and alkaline earths almost exclusively form ionic chemical bonds with nonmetallic elements. Consequently, these elements tend to exist almost entirely as salts or oxides. Only beryllium shows a tendency toward covalent bonding.
11. The oxides and hydroxides of alkali metals tend to be strong bases when dissolved in water.

Alkali and Alkaline Earth Metals provides the reader, whether student or scientist, with an up-to-date understanding regarding each of the elements in these groups—where they came from, how they fit into our current technological society, and where they may lead us.

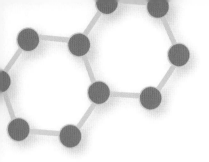

Overview: Chemistry and Physics Background

What *is* an element? To the ancient Greeks, everything on Earth was made from only four elements—earth, air, fire, and water. Celestial bodies—the Sun, moon, planets, and stars—were made of a fifth element: ether. Only gradually did the concept of an element become more specific.

An important observation about nature was that substances can change into other substances. For example, wood burns, producing heat, light, and smoke and leaving ash. Pure metals like gold, copper, silver, iron, and lead can be smelted from their ores. Grape juice can be fermented to make wine and barley fermented to make beer. Food can be cooked; food can also putrefy. The baking of clay converts it into bricks and pottery. These changes are all examples of chemical reactions. Alchemists' careful observations of many chemical reactions greatly helped them to clarify the differences between the most elementary substances ("elements") and combinations of elementary substances ("compounds" or "mixtures").

Elements came to be recognized as simple substances that cannot be decomposed into other even simpler substances by chemical reactions. Some of the elements that had been identified by the Middle Ages are easily recognized in the periodic table because they still have chemical symbols that come from their Latin names. These elements are listed in the following table.

ELEMENTS KNOWN TO ANCIENT PEOPLE

Iron: Fe ("ferrum")	Copper: Cu ("cuprum")
Silver: Ag ("argentum")	Gold: Au ("aurum")
Lead: Pb ("plumbum")	Tin: Sn ("stannum")
Antimony: Sb ("stibium")	Mercury: Hg ("hydrargyrum")
*Sodium: Na ("natrium")	*Potassium: K ("kalium")
Sulfur: S ("sulfur")	

*Sodium and potassium were not isolated as pure elements until the early 1800s, but some of their salts were known to ancient people.

Modern atomic theory began with the work of the English chemist John Dalton in the first decade of the 19th century. As the concept of the atomic composition of matter developed, chemists began to define elements as simple substances that contain only one kind of atom. Because scientists in the 19th century lacked any experimental apparatus capable of probing the structure of atoms, the 19th-century model of the atom was rather simple. Atoms were thought of as small spheres of uniform density; atoms of different elements differed only in their masses. Despite the simplicity of this model of the atom, it was a great step forward in our understanding of the nature of matter. Elements could be defined as simple substances containing only one kind of atom. Compounds are simple substances that contain more than one kind of atom. Because atoms have definite masses, and only whole numbers of atoms can combine to make molecules, the different elements that make up compounds are found in definite proportions by mass. (For example, a molecule of water contains one oxygen atom and two hydrogen atoms, or a mass ratio of oxygen-to-hydrogen of about 8:1.) Since atoms are neither created nor destroyed during ordinary chemical reactions ("ordinary" meaning in contrast to "nuclear" reactions), what happens in chemical reactions is that atoms are rearranged into combinations that differ from the original reactants, but in doing so, the total mass is conserved. Mixtures are combinations of elements that are not in definite proportions. (In salt water, for example, the salt could be 3 percent by mass, or 5 percent by mass, or many other possibilities; regardless of the percentage of salt, it would still be called "salt water.") Chemical reactions are not required to separate the components of mixtures; the components of mixtures can be separated by physical processes such as distillation, evaporation, or precipitation. Examples of elements, compounds, and mixtures are listed in the following table.

The definition of an element became more precise at the dawn of the 20th century with the discovery of the proton. We now know that an atom has a small center called the "nucleus." In the nucleus are one or more protons, positively charged particles, the number of which determine an atom's identity. The number of protons an atom has is referred to as its "atomic number." Hydrogen, the lightest element, has an atomic number of 1, which means each of its atoms contains a single proton. The next element, helium, has an atomic number of 2, which means each of its atoms contain two protons. Lithium has an atomic number of 3, so its atoms have three protons, and so forth, all the way through the periodic table. Atomic nuclei also contain neutrons, but atoms of the same element can have different numbers of neutrons; we call atoms of the same element with different number of neutrons "isotopes."

EXAMPLES OF ELEMENTS, COMPOUNDS, AND MIXTURES

ELEMENTS	COMPOUNDS	MIXTURES
Hydrogen	Water	Salt water
Oxygen	Carbon dioxide	Air
Carbon	Propane	Natural gas
Sodium	Table salt	Salt and pepper
Iron	Hemoglobin	Blood
Silicon	Silicon dioxide	Sand

There are roughly 92 naturally occurring elements—hydrogen through uranium. Of those 92, two elements, technetium (element 43) and promethium (element 61), may once have occurred naturally on Earth, but the atoms that originally occurred on Earth have decayed away, and those two elements are now produced artificially in nuclear reactors. In fact, technetium is produced in significant quantities because of its daily use by hospitals in nuclear medicine. Some of the other first 92 elements—polonium, astatine, and francium, for example—are so radioactive that they exist in only tiny amounts. All of the elements with atomic numbers greater than 92—the so-called transuranium elements—are all produced artificially in nuclear reactors or particle accelerators. As of the writing of this book, the discoveries of the elements through number 118 (with the exception of number 117) have all been reported. The discoveries of elements with atomic numbers greater than 112 have not yet been confirmed, so those elements have not yet been named.

When the Russian chemist Dmitri Mendeleev (1834–1907) developed his version of the periodic table in 1869, he arranged the elements known at that time in order of *atomic mass* or *atomic weight* so that they fell into columns called *groups* or *families* consisting of elements with similar chemical and physical properties. By doing so, the rows exhibit periodic trends in properties going from left to right across the table, hence the reference to rows as *periods* and name "periodic table."

Mendeleev's table was not the first periodic table, nor was Mendeleev the first person to notice *triads* or other groupings of elements with similar properties. What made Mendeleev's table successful and the one we use today are two innovative features. In the 1860s, the concept of *atomic number* had not yet been developed, only the concept of atomic mass. Elements were always listed in order of their atomic masses, beginning with the lightest element, hydrogen, and ending with the heaviest element known at that time, uranium. Gallium and germanium, however, had not yet been discovered. Therefore, if one were listing the known elements in order of atomic mass, arsenic would follow zinc, but that would place arsenic between aluminum and indium. That does not make sense because arsenic's properties are much more like those of phosphorus and antimony, not like those of aluminum and indium.

Russian chemist Dmitri Mendeleev created the periodic table of the elements in 1869. *(Scala/Art Resource)*

To place arsenic in its "proper" position, Mendeleev's first innovation was to leave two blank spaces in the table after zinc. He called the first element *eka-aluminum* and the second element *eka-silicon*, which he said corresponded to elements that had not yet been discovered but whose properties would resemble the properties of aluminum and silicon, respectively. Not only did Mendeleev predict the elements' exis-

Mendeleev's Periodic Table (1871)

Group Period	I	II	III	IV	V	VI	VII	VIII
1	H=1							
2	Li=7	Be=9.4	B=11	C=12	N=14	O=16	F=19	
3	Na=23	Mg=24	Al=27.3	Si=28	P=31	S=32	Cl=35.5	
4	K=39	Ca=40	?=44	Ti=48	V=51	Cr=52	Mn=55	Fe=56, Co=59 Ni=59
5	Cu=63	Zn=65	?=68	?=72	As=75	Se=78	Br=80	
6	Rb=85	Sr=87	?Yt=88	Zr=90	Nb=94	Mo=96	?=100	Ru=104, Rh=104 Pd=106
7	Ag=108	Cd=112	In=113	Sn=118	Sb=122	Te=125	J=127	
8	Cs=133	Ba=137	?Di=138	?Ce=140				
9								
10			?Er=178	?La=180	Ta=182	W=184		Os=195, Ir=197 Pt=198
11	Au=199	Hg=200	Tl=204	Pb=207	Bi=208			
12				Th=231		U=240		

© Infobase Publishing

Dmitri Mendeleev's 1871 periodic table. The elements listed are the ones that were known at that time, arranged in order of increasing relative atomic mass. Mendeleev predicted the existence of elements with masses of 44, 68, and 72. His predictions were later shown to have been correct.

tence, he also estimated what their physical and chemical properties should be in analogy to the elements near them. Shortly afterward, these two elements were discovered and their properties were found to be very close to what Mendeleev had predicted. Eka-aluminum was called *gallium* and eka-silicon was called *germanium*. These discoveries validated the predictive power of Mendeleev's arrangement of the elements and demonstrated that Mendeleev's periodic table could be a predictive tool, not just a compendium of information that people already knew.

The second innovation Mendeleev made involved the relative placement of tellurium and iodine. If the elements are listed in strict order of their atomic masses, then iodine should be placed before tellurium, since iodine is lighter. That would place iodine in a group with sulfur and selenium and tellurium in a group with chlorine and bromine, an arrangement that does not work for either iodine or tellurium. Therefore, Mendeleev rather boldly reversed the order of tellurium and iodine so that tellurium falls below selenium and iodine falls below bromine.

More than 40 years later, after Mendeleev's death, the concept of atomic number was introduced, and it was recognized that elements should be listed in order of atomic number, not atomic mass. Mendeleev's ordering was thus vindicated, since tellurium's atomic number is one less than iodine's atomic number. Before he died, Mendeleev was considered for the Nobel Prize, but did not receive sufficient votes to receive the award despite the importance of his insights.

THE PERIODIC TABLE TODAY

All of the elements in the first 12 groups of the periodic table are referred to as *metals*. The first two groups of elements on the left-hand side of the table are the *alkali metals* and the *alkaline earth* metals. All of the alkali metals are extremely similar to each other in their chemical and physical properties, as, in turn, are all of the alkaline earths to each other. The 10 groups of elements in the middle of the periodic table are *transition metals*. The similarities in these groups are not as strong as those in the first two groups, but still satisfy the general trend of similar chemical and physical properties. The transition metals in the last row are not found in nature but have been synthesized artificially. The metals that follow the transition metals are called *post-transition* metals.

The so-called *rare earth* elements, which are all metals, usually are displayed in a separate block of their own located below the rest of the periodic table. The elements in the first row of rare earths are called *lanthanides* because their properties are extremely similar to the properties of lanthanum. The elements in the second row of rare earths are called *actinides* because their properties are extremely similar to the properties of actinium. The actinides following uranium are called *transuranium* elements and are not found in nature but have been produced artificially.

The far right-hand six groups of the periodic table—the remaining *main group* elements—differ from the first 12 groups in that more than one kind of element is found in them; in this part of the table we find metals, all of the *metalloids* (or *semimetals*), and all of the *nonmetals*. Not counting the artificially synthesized elements in these groups (elements having atomic numbers of 113 and above and that have not yet been named), these six groups contain 7 metals, 8 metalloids, and 16 nonmetals. Except for the last group—the *noble gases*—each individual group has more than just one kind of element. In fact, sometimes nonmetals, metalloids, and metals are all found in the same column, as are the cases with group IVB (C, Si, Ge, Sn, and Pb) and also with group VB (N, P, As, Sb, and Bi). Although similarities in chemical and physical properties are present within a column, the differences are often more striking than the similarities. In some cases, elements in the same column do have very similar chemistry. Triads of such elements include three of the *halogens* in group VIIB—chlorine, bromine, and iodine; and three group VIB elements—sulfur, selenium, and tellurium.

ELEMENTS ARE MADE OF ATOMS

An atom is the fundamental unit of matter. In ordinary chemical reactions, atoms cannot be created or destroyed. Atoms contain smaller *subatomic* particles: protons, neutrons, and electrons. Protons and neutrons are located in the *nucleus,* or center, of the atom and are referred to as *nucleons.* Electrons are located outside the nucleus. Protons and neutrons are comparable in mass and significantly more massive than electrons. Protons carry positive electrical charge. Electrons carry negative charge. Neutrons are electrically neutral.

The identity of an element is determined by the number of protons found in the nucleus of an atom of the element. The number of protons is called an element's atomic number, and is designated by the letter Z. For hydrogen, $Z = 1$, and for helium, $Z = 2$. The heaviest naturally occurring element is uranium, with $Z = 92$. The value of Z is 118 for the heaviest element that has been synthesized artificially.

Atoms of the same element can have varying numbers of neutrons. The number of neutrons is designated by the letter N. Atoms of the same element that have different numbers of neutrons are called *isotopes* of that element. The term *isotope* means that the atoms occupy the same place in the periodic table. The sum of an atom's protons and neutrons is called the atom's *mass number.* Mass numbers are dimensionless whole numbers designated by the letter A and should not be confused with an atom's *mass,* which is a decimal number expressed in units such as grams. Most elements on Earth have more than one isotope. The average mass number of an element's isotopes is called the element's *atomic mass* or *atomic weight.*

The standard notation for designating an atom's atomic and mass numbers is to show the atomic number as a subscript and the mass number as a superscript to the left of the letter representing the element. For example, the two naturally occurring isotopes of hydrogen are written $_1^1H$ and $_1^2H$.

For atoms to be electrically neutral, the number of electrons must equal the number of protons. It is possible, however, for an atom to gain or lose electrons, forming *ions*. Metals tend to lose one or more electrons to form positively charged ions (called *cations*); nonmetals are more likely to gain one or more electrons to form negatively charged ions (called *anions*). Ionic charges are designated with superscripts. For example, a calcium ion is written as Ca^{2+}; a chloride ion is written as Cl^-.

THE PATTERN OF ELECTRONS IN AN ATOM

During the 19th century, when Mendeleev was developing his periodic table, the only property that was known to distinguish an atom of one element from an atom of another element was relative mass. Knowledge of atomic mass, however, did not suggest any relationship between an element's mass and its properties. It took several discoveries—among them that of the electron in 1897 by the British physicist John Joseph ("J. J.") Thomson, *quanta* in 1900 by the German physicist Max Planck, the wave nature of matter in 1923 by the French physicist Louis de Broglie, and the mathematical formulation of the quantum mechanical model of the atom in 1926 by the German physicists Werner Heisenberg and Erwin Schrödinger (all of whom collectively illustrate the international nature of science)—to elucidate the relationship between the structures of atoms and the properties of elements.

The number of protons in the nucleus of an atom defines the identity of that element. Since the number of electrons in a neutral atom is equal to the number of protons, an element's atomic number also reveals how many electrons are in that element's atoms. The electrons occupy regions of space that chemists and physicists call *shells*. The

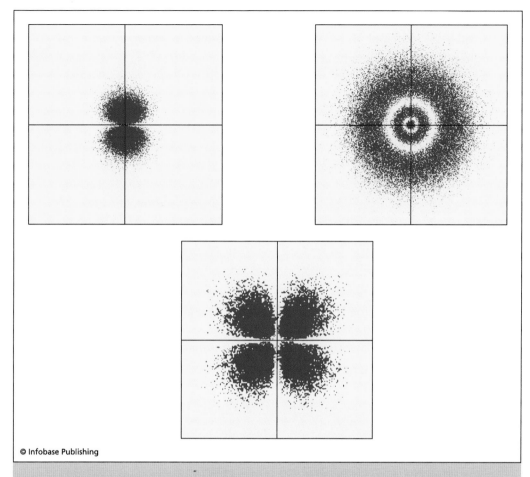

Hydrogen wave-function distributions for electrons in various excited states take on widely varying configurations.

shells are further divided into regions of space called *subshells*. Subshells are related to angular momentum, which designates the shape of the electron orbit space around the nucleus. Shells are numbered 1, 2, 3, 4, and so forth (in theory out to infinity). In addition, shells may be designated by letters: The first shell is the "K-shell," the second shell the "L-shell," the third the "M-shell," and so forth. Subshells have letter designations, "s," "p," "d," and "f" being the most common. The *n*th shell has *n* possible subshells. Therefore, the first shell has only an "s" subshell, designated "1s"; the second shell has both "s" and "p" subshells

("2s" and "2p"); the third shell "3s," "3p," and "3d"; and the fourth shell "4s," "4p," "4d," and "4f." (This pattern continues for higher-numbered shells, but this is enough for now.)

An "s" subshell is spherically symmetric and can hold a maximum of 2 electrons. A "p" subshell is dumbbell-shaped and holds 6 electrons, a "d" subshell 10 electrons, and an "f" subshell 14 electrons, with increasingly complicated shapes.

As the number of electrons in an atom increases, so does the number of shells occupied by electrons. In addition, because electrons are all negatively charged and tend to repel each other *electrostatically,* as the number of the shell increases, the size of the shell increases, which means that electrons in higher-numbered shells are located, on the average, farther from the nucleus. Inner shells tend to be fully occupied with the maximum number of electrons they can hold. The electrons in the outermost shell, which is likely to be only partially occupied, will determine that atom's properties.

Physicists and chemists use *electronic configurations* to designate which subshells in an atom are occupied by electrons as well as how many electrons are in each subshell. For example, nitrogen is element number 7, so it has seven electrons. Nitrogen's electronic configuration is $1s^2 2s^2 2p^3$; a superscript designates the number of electrons that occupy a subshell. The first shell is fully occupied with its maximum of two electrons. The second shell can hold a maximum of eight electrons, but it is only partially occupied with just five electrons—two in the 2s subshell and three in the 2p. Those five outer electrons determine nitrogen's properties. For a heavy element like tin (Sn), electronic configurations can be quite complex. Tin's configuration is $1s^2 2s^2 2p^6 3s^2 3p^6 4s^2 3d^{10} 4p^6 5s^2 4d^{10} 5p^2$ but is more commonly written in the shorthand notation [Kr] $5s^2 4d^{10} 5p^2$ where [Kr] represents the electron configuration pattern for the noble gas Krypton. (The pattern continues in this way for shells with higher numbers.) The important thing to notice about tin's configuration is that all of the shells except the last one are fully occupied. The fifth shell can hold 32 electrons, but in tin there are only four electrons in the fifth shell. The outer electrons determine an element's properties. The following table illustrates the electronic configurations for nitrogen and tin.

ELECTRONIC CONFIGURATIONS FOR NITROGEN AND TIN

ELECTRONIC CONFIGURATION OF NITROGEN (7 ELECTRONS)

Energy Level	Shell	Subshell	Number of Electrons
1	K	s	2
2	L	s	2
		p	3
			7

ELECTRONIC CONFIGURATION OF TIN (50 ELECTRONS)

Energy Level	Shell	Subshell	Number of Electrons
1	K	s	2
2	L	s	2
		p	6
3	M	s	2
		p	6
		d	10
4	N	s	2
		p	6
		d	10
5	O	s	2
		p	2
			50

ATOMS ARE HELD TOGETHER BY CHEMICAL BONDS

Fundamentally, a *chemical bond* involves either the sharing of two electrons or the transfer of one or more electrons to form ions. Two atoms of nonmetals tend to share pairs of electrons in what is called a *covalent* bond. By sharing electrons, the atoms remain more or less electrically neutral. However, when an atom of a metal approaches an atom of a nonmetal, the more likely event is the transfer of one or more electrons from the metal atom to the nonmetal atom. The metal atom becomes a positively charged ion and the nonmetal atom becomes a negatively charged ion. The attraction between opposite charges provides the force that holds the atoms together in what is called an *ionic* bond. Many chemical bonds are also intermediate in nature between covalent and ionic bonds and have characteristics of both types of bonds.

IN CHEMICAL REACTIONS, ATOMS REARRANGE TO FORM NEW COMPOUNDS

When a substance undergoes a *physical change,* the substance's name does not change. What may change is its temperature, its length, its *physical state* (whether it is a solid, liquid, or gas), or some other characteristic, but it is still the same substance. On the other hand, when a substance undergoes a *chemical change,* its name changes; it is a different substance. For example, water can decompose into hydrogen gas and oxygen gas, each of which has substantially different properties from water, even though water is composed of hydrogen and oxygen atoms.

In chemical reactions, the atoms themselves are not changed. Elements (like hydrogen and oxygen) may combine to form compounds (like water), or compounds can be decomposed into their elements. The atoms in compounds can be rearranged to form new compounds whose names and properties are different from the original compounds. Chemical reactions are indicated by writing chemical equations such as the equation showing the decomposition of water into hydrogen and oxygen: $2 H_2O \ (l) \rightarrow 2 H_2 \ (g) + O_2 \ (g)$. The arrow indicates the direction in which the reaction proceeds. The reaction begins with the *reactants* on the left and ends with the *products* on the right. We sometimes designate the physical state of a reactant or product in parentheses—"s" for solid, "*l*" for liquid, "g" for gas, and "aq" for *aqueous* solution (in other words, a solution in which water is the *solvent*).

THE NUCLEI OF ATOMS CHANGE IN NUCLEAR REACTIONS

In ordinary chemical reactions, chemical bonds in the reactant species are broken, the atoms rearrange, and new chemical bonds are formed in the product species. These changes only affect an atom's electrons; there is no change to the nucleus. Hence there is no change in an element's identity. On the other hand, nuclear reactions refer to changes in an atom's nucleus (whether or not there are electrons attached). In most nuclear reactions, the number of protons in the nucleus changes, which means that elements are changed, or *transmuted,* into different elements. There are several ways in which transmutation can occur. Some transmutations occur naturally, while others only occur artificially in nuclear reactors or particle accelerators.

The most familiar form of transmutation is *radioactive decay,* a natural process in which a nucleus emits a small particle or *photon* of light. Three common modes of decay are labeled *alpha, beta,* and *gamma* (the first three letters of the Greek alphabet). Alpha decay occurs among elements at the heavy end of the periodic table, basically elements heavier than lead. An alpha particle is a nucleus of helium 4 and is symbolized as $^{4}_{2}He$ or α. An example of alpha decay occurs when uranium 238 emits an alpha particle and is changed into thorium 234 as in the following reaction: $^{238}_{92}U \rightarrow \, ^{4}_{2}He + \, ^{234}_{90}Th$. Notice that the *parent* isotope, U-238, has 92 protons, while the *daughter* isotope, Th-234, has only 90 protons. The decrease in the number of protons means a change in the identity of the element. The mass number also decreases.

Any element in the periodic table can undergo beta decay. A beta particle is an electron, commonly symbolized as β^- or e^-. An example of beta decay is the conversion of cobalt 60 into nickel 60 by the following reaction: $^{60}_{27}Co \rightarrow \, ^{60}_{28}Ni + e^-$. The atomic number of the daughter isotope is one greater than that of the parent isotope, which maintains charge balance. The mass number, however, does not change.

In gamma decay, photons of light (symbolized by γ) are emitted. Gamma radiation is a high-*energy* form of light. Light carries neither mass nor charge, so the isotope undergoing decay does not change identity; it only changes its energy state.

Elements also are transmuted into other elements by nuclear *fission* and *fusion*. Fission is the breakup of very large nuclei (at least as heavy as uranium) into smaller nuclei, as in the fission of U-236 in the following reaction: $^{236}_{92}U \rightarrow {}^{94}_{36}Kr + {}^{139}_{56}Ba + 3n$, where n is the symbol for a neutron (charge = 0, mass number = +1). In fusion, nuclei combine to form larger nuclei, as in the fusion of hydrogen isotopes to make helium. Energy may also be released during both fission and fusion. These events may occur naturally—fusion is the process that powers the Sun and all other stars—or they may be made to occur artificially.

Elements can be transmuted artificially by bombarding heavy target nuclei with lighter projectile nuclei in reactors or accelerators. The transuranium elements have been produced that way. Curium, for example, can be made by bombarding plutonium with alpha particles. Because the projectile and target nuclei both carry positive charges, projectiles must be accelerated to velocities close to the speed of light to overcome the force of repulsion between them. The production of successively heavier nuclei requires more and more energy. Usually, only a few atoms at a time are produced.

ELEMENTS OCCUR WITH DIFFERENT RELATIVE ABUNDANCES

Hydrogen overwhelmingly is the most abundant element in the universe. Stars are composed mostly of hydrogen, followed by helium and only very small amounts of any other element. Relative abundances of elements can be expressed in parts per million, either by mass or by numbers of atoms.

On Earth, elements may be found in the *lithosphere* (the rocky, solid part of Earth), the *hydrosphere* (the aqueous, or watery, part of Earth), or the atmosphere. Elements such as the noble gases, the rare earths, and commercially valuable metals like silver and gold occur in only trace quantities. Others, like oxygen, silicon, aluminum, iron, calcium, sodium, hydrogen, sulfur, and carbon are abundant.

HOW NATURALLY OCCURRING ELEMENTS HAVE BEEN DISCOVERED

For the elements that occur on Earth, methods of discovery have been varied. Some elements—like copper, silver, gold, tin, and lead—have been known and used since ancient or even prehistoric times. The origins of their early *metallurgy* are unknown. Some elements, like phosphorus, were discovered during the Middle Ages by alchemists who recognized that some mineral had an unknown composition. Sometimes, as in the case of oxygen, the discovery was by accident. In other instances—as in the discoveries of the alkali metals, alkaline earths, and lanthanides—chemists had a fairly good idea of what they were looking for and were able to isolate and identify the elements quite deliberately.

To establish that a new element has been discovered, a sample of the element must be isolated in pure form and subjected to various chemical and physical tests. If the tests indicate properties unknown in any other element, it is a reasonable conclusion that a new element has been discovered. Sometimes there are hazards associated with isolating a substance whose properties are unknown. The new element could be toxic, or so reactive that it can explode, or extremely radioactive. During the course of history, attempts to isolate new elements or compounds have resulted in more than just a few deaths.

HOW NEW ELEMENTS ARE MADE

Some elements do not occur naturally, but can be synthesized. They can be produced in nuclear reactors, from collisions in particle accelerators, or can be part of the *fallout* from nuclear explosions. One of the elements most commonly made in nuclear reactors is technetium. Relatively large quantities are made every day for applications in nuclear medicine. Sometimes, the initial product made in an accelerator is a heavy element whose atoms have very short *half-lives* and undergo radioactive decay. When the atoms decay, atoms of elements lighter than the parent atoms are produced. By identifying the daughter atoms, scientists can work backward and correctly identify the parent atoms from which they came.

The major difficulty with synthesizing heavy elements is the number of protons in their nuclei (Z > 92). The large amount of positive charge makes the nuclei unstable so that they tend to disintegrate either by radioactive decay or *spontaneous fission*. Therefore, with the exception of a few transuranium elements like plutonium (Pu) and americium (Am), most artificial elements are made only a few atoms at a time and so far have no practical or commercial uses.

THE ALKALI AND ALKALINE EARTH METALS SECTION OF THE PERIODIC TABLE

Comprising the left-hand column of the periodic table, after the element of hydrogen, are the following alkali metals:

- lithium,
- sodium,
- potassium,
- rubidium,
- cesium, and
- francium.

Comprising the adjoining column of the periodic table are the following alkaline earth metals:

- beryllium,
- magnesium,
- calcium,
- strontium,
- barium, and
- radium.

Since the alkali metals all have just one electron in their outer shells and the alkaline earth metals have two, they are located on the left edge of the periodic table on page 124 in the first and second columns. The following is the key to understanding each element's information box that appears at the beginning of each chapter.

Having an unfilled outer electron shell allows an element to react easily with others that can accommodate another electron in an outer shell. Therefore the alkali and alkaline earth metals are chemically quite active and are not found in the pure elemental form on Earth.

	Element	
K		M.P.°
L	**E**$_Z$	B.P.°
M		C.P.°
N		
O		
P	Oxidation states	
Q	Atomic weight	
	Abundance%	

Information box key. E represents the element's letter notation (for example, H = hydrogen), with the Z subscript indicating proton number. Orbital shell notations appear in the column on the left. For elements that are not naturally abundant, the mass number of the longest-lived isotope is given in brackets. The abundances (atomic %) are based on meteorite and solar wind data. The melting point (M.P.), boiling point (B.P.), and critical point (C.P.) temperatures are expressed in Celsius. Sublimation and critical temperatures are indicated by s and t.

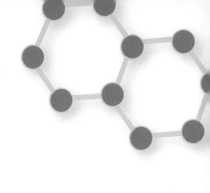

PART

I

Alkali Metals

INTRODUCTION TO ALKALI METALS

Several trends in properties of the alkali metals occur as the atomic mass increases. Lithium has a much higher specific heat than the other alkalis; specific heats decrease upon descending the column. Lithium is the lightest of the alkalis; density increases going down the column. *Electron affinities* decrease as atomic weight increases because the influence of the nuclear charge on the outermost—or valence—electron lessens due to screening by the core electrons. All alkali metals react readily with water. However, the reaction with water becomes more violent as the atomic weight of the alkali increases.

The chemistry of the alkali metals is determined primarily by the anions with which they bond. Differences in behavior are due mostly to variations in sizes of ions and to different *heats of hydration*. A large variety of inorganic compounds have been made with alkali metals, including *hydrides, hydroxides, nitrates, nitrides, oxides,* and *peroxides, permanganates, phosphates,* and *silicates.* In addition, a large number of *organo-alkali compounds* have been prepared in which, most commonly, sodium or potassium combine with various *hydrocarbons.*

Analytical chemistry of the alkalis is difficult. There are a few complex reagents that will *precipitate* with ions like Na^+ and K^+, but the reactions are best done using ether or alcohol as the solvent in place of water. It is more common to detect the presence of alkali metals in solutions or compounds by the characteristic colors the alkali metal ions impart to flames. In descending order, lithium salts give a carmine color, sodium salts a yellow color, potassium a violet color, rubidium a bluish red color, and cesium a blue color.

THE DISCOVERY AND NAMING OF ALKALI METALS

Sodium and potassium salts were known to ancient people. For example, in the Hebrew story of Lot and his family fleeing the destruction of the cities of Sodom and Gomorrah (Genesis, chapter 19, verse

Lithium is a soft, silvery-white, solid element. *(Martyn F. Chillmaid/ Photo Researchers, Inc.)*

26), it is recorded that "Lot's wife . . . looked back and became a pillar of salt." The symbol Na for sodium comes from the Latin name for sodium, *natrium.* The symbol K for potassium comes from its Latin name, *kalium.*

In the first decade of the 1800s, Sir Humphrey Davy established himself as a careful experimenter in the field of *electrochemistry.* In 1807, in his first attempts to decompose *soda ash* (NaOH) and *potash* (KOH), he passed electrical currents through aqueous solutions containing these compounds. However, he succeeded only in decomposing the water into hydrogen and oxygen gases. Davy then passed a current through a slightly moistened piece of potash. He observed gas bubbles at the positive electrode (the *anode*). At the negative electrode (the *cathode*), he observed the formation of globules whose appearance resembled that of mercury. Some of the globules burst into flame and exploded (they also burst into flame when thrown into water). Other globules were quickly tarnished. Davy discovered that the metal reacted with water to liberate hydrogen gas; it was the burning of the hydrogen that produced the flame. Because Davy had obtained this new metal from potash, he named it "potassium."

Davy then repeated the experiment using soda ash. He found that a stronger electrical current was required, and within a few days of his isolation of potassium, he discovered sodium (named because it came from soda ash). At first, other scientists doubted that Davy's potassium and sodium were, in fact, true elements. They thought that maybe Davy's "elements" were compounds of potassium or sodium and hydrogen. Davy was able, however, to demonstrate successfully their elemental nature.

One of the greatest chemists of the 19th century was Jöns Jakob Berzelius (1779–1848) of Sweden. In 1817, Berzelius put one of his assistants, Swedish chemist Johan August Arfwedson (1792–1841), to work analyzing the mineral petalite. Arfwedson could account for 96 percent of petalite's content, but the remaining 4 percent was a mystery. By 1818, Berzelius and Arfwedson had concluded that petalite must contain an unknown alkali metal. Petalite's composition proved to be lithium aluminum silicate, and Arfwedson is acknowledged as lithium's discoverer. The name "lithium" comes from the Greek word *lithos,* which means "stone." That same year, the red color that lithium salts impart to flames was discovered. Arfwedson tried unsuccessfully to obtain pure lithium metal by *electrolysis;* it was W. T. Brande and Davy who independently finally succeeded in isolating white lithium metal by the electrolysis of lithia. For many years, experimenters were able to obtain only tiny quantities of lithium. It was not until 1855 that appreciable quantities of lithium were obtained by the electrolysis of molten lithium chloride (LiCl).

It was 42 years after the discovery of lithium that the next alkali metal, cesium, was discovered. During the decade of the 1850s, German chemist Robert Wilhelm Bunsen (1811–99) and German physicist Gustav Robert Kirchhoff (1824–87) had developed the Kirchhoff-Bunsen *spectroscope,* which they used to determine the chemical compositions of minerals. Minerals can be heated until they radiate light. Bunsen, in fact, had invented a burner (the so-called Bunsen burner) in 1854–55 whose flame itself is colorless. A spectroscope uses a prism to separate the emitted light into its colors, which appear as bright lines against a dark background. Since each element has its own distinctive spectral pattern (much like the unique fingerprints of humans), the elements in a mineral can be identified. In 1860, Bunsen and Kirchhoff detected new blue spectral lines that did not correspond to any known element. They determined the new element to be an alkali metal and named it "cesium" after the Greek word *caesius,* which refers to the blue color of the upper atmosphere. It took 20 years to isolate cesium metal, but that feat was finally accomplished in Bunsen's laboratory through the electrolysis of cesium cyanide.

In 1861, shortly after their announcement of the discovery of cesium, Bunsen and Kirchhoff announced the discovery of rubidium in the mineral lepidolite. Rubidium's presence was indicated by two previously unknown deep red lines in the mineral's spectrum. The name "rubidium" comes from the Latin word *rubidus,* which refers to red of the deepest color. Bunsen succeeded in isolating rubidium metal itself.

Bunsen was extremely famous in his own time and revered by his students for his part in the development of the technique of spectral analysis; numerous other elements were later discovered by chemists using Bunsen and Kirchoff's spectroscope. Those elements include gallium, helium, holmium, indium, lutetium, neodymium, praseodymium, thallium, thulium, and ytterbium.

The last discovery of an alkali metal occurred almost 80 years later. In 1939, Parisian physicist Marguerite Perey (1909–75) observed an unusual rate of radioactive decay in a sample of a salt of actinium (element 89). She managed to isolate the new element, showed that it was an alkali metal, and named it "francium" in honor of her native country, France. Because francium's longest-lived isotope has a half-life of only 21 minutes, francium is the rarest element below element 98 in the periodic table, which explains why francium was discovered much later than the other radioactive elements in that part of the table.

1

Lithium

Lithium is element number 3, a soft, silvery-white, solid element with a density of 0.535 g/cm³. It is the lightest of all metals and can be *alloyed* with other metals to make lightweight structural materials.

Lithium occurs in *silicate* ores (minerals that principally contain silicon and oxygen) in association with aluminum and the other alkalis: The principal ore is spodumene ($LiAlSi_2O_6$). Lithium mining occurs mostly in the United States, Russia, China, and Australia, and in addition, it is extracted from salt deposits in Chile.

Lithium is an extremely *electropositive* element and has the highest *oxidation potential* of all elements. Consequently, the metal is an extremely good *reducing agent*. Like all of the alkali metals, compounds of lithium contain lithium only as the +1 ion.

THE BASICS OF LITHIUM

Symbol: Li
Atomic number: 3
Atomic mass: 6.941
Electronic configuration: $1s^2 2s^1$

T_{melt} = 358°F (181°C)
T_{boil} = 2,448°F (1,342°C)

Abundance:
In Earth's crust 18 ppm
In seawater 0.2 ppm

	Lithium	
2		180.5°
1	**Li** 3	1342°
	+1	
	6.941	
	1.86×10^{-7}%	

Isotope	Z	N	Relative Abundance
$^6_3$Li	3	3	7.59%
$^7_3$Li	3	4	92.41%

This chapter discusses the synthesis of lithium in stars, the chemistry of lithium, and several important modern-day applications and uses of lithium.

THE ASTROPHYSICS OF LITHIUM

Lithium is one of the heaviest of nuclei produced during the first few minutes of big-bang nucleosynthesis, formed either by fusion of deuterium and tritium or by fusion of two alpha particles as follows:

$$2\,^2_1H + \,^3_1H \rightarrow \,^7_3Li$$

$$^4_2He + \,^4_2He \rightarrow \,^7_3Li + \,^1_1p.$$

The same reactions in stellar cores produce lithium nuclei, but these are continually suffering bombardment by high-energy protons. A temperature of only a few million kelvins, which is exceeded in all stellar interiors, is sufficient to destroy any lithium that is produced. Therefore, observations of the element in the spectra of certain stellar types or in the interstellar medium are of great interest to astronomers.

Lithium has two stable isotopes, ^{6}Li and ^{7}Li, both of which can serve as indicators of important astrophysical processes. In particular, the ratio of these isotopes can provide clues about interstellar production of the light elements as well as information on convection in stellar atmospheres.

In the space between the stars—the interstellar medium (ISM)—a very different process is responsible for lithium production. In this case, *spallation,* typically defined as the breaking apart of a substance, refers to the breaking up of larger interstellar nuclei (C, N, O) into the smaller nuclei of lithium 6 or lithium 7, which occurs by collision with sufficiently energetic cosmic ray protons. The same collisional reactions can take place in the outer atmospheres of stars with strong magnetic fields or in regions that experience thermonuclear or gravitational shocks, which may explain the observations of lithium spectra in varying stellar types, including chemically peculiar and metal-poor halo stars.

The ^{6}Li/^{7}Li abundance in stars varies widely, and the production mechanisms are far from well understood. Current research examines the relative roles of *convection, diffusion,* and spallation to gain a greater understanding of *stellar evolution.*

THE CHEMISTRY OF LITHIUM

The most important chemical property of all the alkali metals, including lithium, is that they are extremely chemically reactive. For example, they all react vigorously with water, in the process forming hydrogen gas, shown in the following reaction between lithium and water:

$$2 \text{ Li (s)} + 2 \text{ H}_2\text{O } (l) \rightarrow 2 \text{ LiOH (aq)} + \text{H}_2 \text{ (g)}.$$

This reaction is not sufficiently *exothermic* to ignite the hydrogen. Because of lithium's reactivity with water, however, any samples of lithium metal must be packaged so as to exclude them from water and from the moisture in air.

Any compound that lithium forms will be an ionic compound, with lithium present as the lithium ion (Li$^+$). In addition to its reaction with water, lithium also reacts readily with oxygen gas in the atmosphere to

(continued on page 11)

FUEL FOR FUSION

While a deuterium-tritium (D-T) mix is the fuel of choice for sustained fusion research, tritium on Earth is scarce. Produced by cosmic ray protons colliding with nitrogen in the upper atmosphere, trace amounts are found in air and less abundantly in water. That is where lithium comes in. Bombarding lithium 6 or lithium 7 atoms with high-energy neutrons results in atoms of tritium and helium.

$$^6Li + n \rightarrow {}^3H + {}^4He + 4.78 \text{ MeV}$$

$$^7Li + n + 2.47 \text{ MeV} \rightarrow {}^3H + {}^4He + n$$

Fortunately, lithium is found with a higher abundance than lead in Earth's crust and to a lesser extent in seawater.

(continues)

Fusion research relies on the bombardment of lithium with neutrons to produce tritium. Pictured here is the interior of the Joint European Torus device at Culham, Oxfordshire, England: It is part of the European Fusion Development Agreement initiative to research and investigate the viability of fusion power. *(EFDA-JET/Photo Researchers, Inc.)*

(continued)

Fusion reactors, as well as thermonuclear weapons, produce fast neutrons in profusion as a result of D-D and D-T fusion:

$$D + T \rightarrow {}^4He + n + 17.588 \text{ MeV}$$

$$D + D \rightarrow {}^3He + n + 3.268 \text{ MeV.}$$

Because the neutrons are so much lighter than the tritium or helium nuclei produced, they carry off most of the energy and easily stimulate the lithium fission processes shown above.

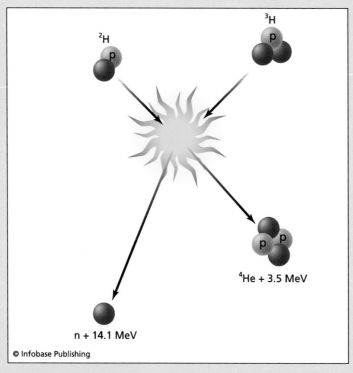

© Infobase Publishing

When deuterium (^{2}H) fuses with tritium (^{3}H), high-energy neutrons are produced.

A particular benefit of lithium in thermonuclear weapons comes as a result of its stable nature. Pure deuterium is difficult to store, as it must be kept at very low temperatures. When combined with lithium, however, into lithium deuteride (LiD)—a solid with low reactivity—deuterium can be easily transported at ambient temperatures. The lithium itself is usually enriched with the lithium 6 isotope when used for this purpose.

(continued from page 8)

form lithium oxide (Li_2O) and with chlorides, sulfates, and carbonates to form lithium chloride (LiCl), lithium sulfate (Li_2SO_4), and lithium carbonate (Li_2CO_3), respectively. Lithium can also react with hydrogen gas to form lithium hydride (LiH), an ionic compound in which the hydrogen is present as a −1 ion instead of its usual +1 ion. As a solid with a high density of hydrogen, lithium hydride shows promise as a means of storing hydrogen for use in *fuel cells*. In application, adding a little water will release the hydrogen, as demonstrated by the following equation:

$$LiH\ (s) + H_2O\ (l) \rightarrow LiOH\ (aq) + H_2\ (g).$$

LITHIUM AND BIPOLAR DISORDER

Bipolar disorder (formerly referred to as "manic-depressive disorder") is characterized by extreme mood swings. During a manic episode, the sufferer tends to engage in risky behaviors, becomes highly excitable, talks incessantly, and exhibits little need for sleep. During a depressive episode, the tendency is to sleep poorly, undergo extremely low self-esteem, and have difficulty concentrating. A direst possibility during the depressive stage is to attempt suicide.

In the 1940s, scientists discovered that lithium carbonate helps to even out a person's high and low mood swings. By the 1970s, lithium was being prescribed quite commonly for patients diagnosed with bipolar disorder. More than 50 years later, scientists are finally beginning to understand how lithium is able to regulate the radical nature of bipolar emotionality. Experiments on mice and monkeys indicate that lithium can regulate the uptake of glutamate—a neurotransmitter that facili-

HOW LITHIUM CAN ALLEVIATE EXCESS CO_2

Fears of global warming are prompting many scientists to research ways of eliminating CO_2 from exhaust products. The following method has been known for decades. If water vapor is present, as it most often is in situations on Earth, lithium hydroxide (LiOH) can absorb carbon dioxide gas, embedding it in solid lithium carbonate (Li_2CO_3) via these following two steps:

$$LiOH \ (s) + H_2O \ (g) \rightarrow LiOH \times H_2O \ (s)$$

$$2 \ LiOH \times H_2O \ (s) + CO_2 \ (g) \rightarrow Li_2CO_3 \ (s) + 3H_2O \ (g).$$

This convenient method has been used successfully to keep air breathable for humans living in extreme environments like deep-sea diving apparatuses, submarines, space suits, and spacecraft. One of the many serious problems that occurred during the *Apollo 13* lunar mission involved LiOH canisters that didn't fit the air purification system.

Lithium hydroxide's light weight and relatively low cost have recommended it over other alkali hydroxides, any of which could do the same job. Lithium hydroxide is water soluble, however, which is undesirable in situations where less care might be taken to keep it dry. The Toshiba company has been working on a lithium silicate absorption device that could turn out to be cost-effective for use in absorbing CO_2 from car exhaust. The difficulty is in making a device small enough to fit in a typical automobile exhaust system.

tates message transmission to neurons (synaptic connections) in the brain. Reuptake of glutamate is an important factor, and lithium seems to be able to adjust the limits of glutamate uptake or retention, providing bounds that keep a patient from experiencing extremes. The range of glutamate for normal behavior is apparently produced naturally in people without bipolar disorder. According to H. J. Kim of the University of Minnesota Medical Department, "Increased synaptic connections may underlie the mood-stabilizing effects of lithium in patients with bipolar disorder and could contribute to the convulsions produced by excessive doses of this drug."

Lithium carbonate can be taken in various forms—as a tablet, a capsule, or a liquid—with a usual dose of 300 mg normally taken two to four times daily. The possible side effects may include the following: water retention, the need for frequent urination, weight gain, nausea, or shaky hands. Adjusting the dosage can control the frequency of such side effects. Although there are newer and more expensive antidepressive drugs on today's market, lithium remains the safest and most tested drug for treating bipolar disorder.

TECHNOLOGY AND CURRENT USES OF LITHIUM

Lithium is one of the most versatile alkali metals. From batteries to lubricants and catalysts to pharmaceuticals, lithium and its compounds find a wide variety of uses. Lithium stands out among the alkali metals for several reasons. It is the lightest alkali metal and the least-dense element that is solid at room temperature and atmospheric pressure. Lithium has the highest melting point of the alkali metals. Its ion (Li^+) is the smallest ion in the family. Although less abundant that sodium or potassium, lithium is still relatively inexpensive to obtain.

Lithium-ion batteries are compact and lightweight, and have high energy density and long lifetimes. Early batteries were hazardous because lithium metal reacts explosively with water. That problem was solved, however, when substances were found that could be combined with lithium to make it impervious to water. Lithium batteries today have major uses that include the following electronic applications: rechargeable and nonrechargeable batteries, battery packs, chargers, camera flashes, power supplies for communications, amplifiers, mobile phones,

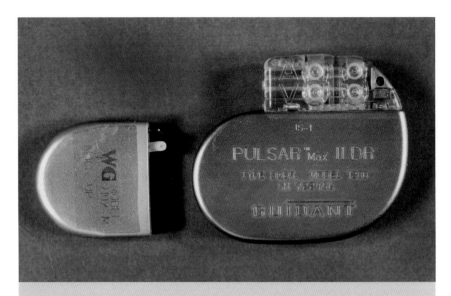

Lithium-ion batteries are compact and lightweight and have high energy density and long lifetimes, which make them useful in pacemakers like the Guidant Pulsar dual-chamber pacemaker pictured here with an eight-year lithium-oxide battery at left. *(Leonard Lessin/Photo Researchers, Inc.)*

laptop computers, wristwatches, standby power supplies, electric razors and toothbrushes, calculators, portable radios and televisions, personal digital assistants, implantable medical devices, and toys. Lithium batteries are especially desirable in applications such as pacemakers, where the battery must be small and have a very long lifetime.

In industry, a major use of lithium is in lubricating greases. Lubricating greases are made with a base consisting of a mineral or synthetic oil. A metallic soap is added as a thickener. In this case, the soap is lithium stearate ($LiC_{18}H_{35}O_2$). Lithium stearate grease can be purchased at hardware stores and home improvement centers and works well as a lubricant in situations where metal parts tend to rub together.

Elemental lithium is a *catalyst*. Catalysts are substances that speed up chemical reactions without themselves being consumed during the reaction. In pharmaceutical syntheses, an example is the synthesis of methamphetamine from ephedrine. In the polymer industry, lithium is used to catalyze the manufacture of ethane (C_2H_6) from ethylene (C_2H_4), as shown in the following reaction:

$$CH_2 = CH_2 \text{ (g)} + H_2 \text{ (g)} \xrightarrow{\text{Li}} CH_3 - CH_3 \text{ (g)}.$$

Other uses for the metal include lightweight alloys (such as aluminum-lithium aircraft parts) and glass for television tubes. Several lithium compounds have important applications in glass and porcelain manufacturing processes. The use of lithium carbonate (Li_2CO_3) to treat bipolar disorder has already been discussed. Lithium carbonate is also used as a *flux* in the smelting of aluminum. Lithium chloride (LiCl) is very water-absorbent and is used as a *desiccant* in air conditioning systems in submarines. Lithium hydride (LiH) can be used to store hydrogen atoms that can later be released as hydrogen gas. Lithium hydroxide (LiOH) is used as an absorbent to remove carbon dioxide.

A nuclear application of lithium is in thermonuclear weapons and fusion research. In a weapon or fusion reactor, nuclear fusion occurs between two isotopes of hydrogen—deuterium and tritium. Deuterium occurs naturally and has an abundant supply in the world's oceans (it is present in about 0.015 percent of water molecules). Tritium, on the other hand, is radioactive, has a relatively short half-life, and does not occur naturally. Tritium can be manufactured, however, by bombarding lithium 6 with neutrons.

Worldwide lithium production is about 12,500 metric tons per year. With increasing demand for lightweight batteries and power sources that do not depend on fossil fuels, the use of lithium will almost certainly continue to grow.

2

Sodium

Sodium, element number 11, is a soft and silvery gray metal with a density of 0.97 g/cm³. Since this value is less than the density of water, sodium metal floats (although it reacts violently with the water upon contact). Sodium is so soft that it can be cut and shaped with a butter knife. It is the most abundant of the alkali metals, readily obtained from seawater, and used in a large number of applications. Of all the alkali metals, sodium metal is the one produced and used in industry in greatest quantity.

Sodium is most familiar to people as the source of "saltiness" in table salt, a valuable commodity in trade and commerce for millennia. The most abundant mineral that contains sodium chloride is halite, or "rock salt." Rock salt deposits when inland salty seas evaporate, leaving the salt behind. (Examples of such bodies of water in the United States

THE BASICS OF SODIUM

Symbol: Na
Atomic number: 11
Atomic mass: 22.989770
Electronic configuration: [Ne]3s^1

T_{melt} = 208°F (98°C)
T_{boil} = 1,621°F (883°C)

Abundances:
In Earth's crust 22,700 ppm
In seawater 10,600 ppm

	Sodium	
2		97.80°
8	**Na**$_{11}$	883°
1		
	+1	
	22.989770	
	0.000187%	

Isotope	Z	N	Relative Abundance
$^{23}_{11}$Na	11	12	100%

are the Salton Sea in southeastern California, Mono Lake in eastern California, and the Great Salt Lake in northern Utah. A famous Old World example is the Dead Sea in Israel.)

In this chapter, the reader will learn about the production of sodium in stars, the chemistry of sodium, the role of sodium in food, and useful applications of sodium.

THE ASTROPHYSICS OF SODIUM

Astrophysical sodium is produced in various ways that depend on the mass and life stage of a given star. In very massive stars that experience carbon-burning, sodium 23 results from the following reaction and is then distributed into the interstellar medium (ISM) in the supernova explosion of the star.

$$^{12}_{6}C + ^{12}_{6}C \rightarrow ^{23}_{11}Na + ^{1}_{1}p$$

Other interesting mechanisms appear to be at work to produce sodium in the atmospheres of red giant stars, where the abundance indicates that some poorly understood mixing in the atmosphere must take place. Additionally, yellow supergiant stars show a temperature-dependent sodium excess over expected abundance.

The sodium abundance in the ISM is also higher than expected from what would be produced by supernovae alone, especially in the content of interplanetary dust particles. It may be that stellar winds and *planetary nebulae* play a large role in delivering sodium into the universe.

Continuing research into how sodium is produced and delivered throughout interstellar space will help astrophysicists fine-tune their understanding of stellar processes.

THE CHEMISTRY OF SODIUM

Sodium metal reacts vigorously with water, as shown in the following chemical equation:

$$2\,Na\,(s) + 2\,H_2O\,(l) \rightarrow 2\,NaOH\,(aq) + H_2\,(g).$$

The reaction is so exothermic that the heat produced can ignite the hydrogen gas and cause it to explode. The demonstration of this reaction is common in chemistry classrooms. The demonstration must be performed very carefully under supervision so that the container in which the reaction is taking place does not explode, spewing hot NaOH (a *caustic* alkali) on nearby persons. As a safety measure, the demonstration is done with only tiny pellets of sodium.

There are a number of important compounds that contain sodium as the Na^+ ion. (Like all the alkali metal ions, the +1 ion is the only common ion.) Rock salt—or table salt in crystalline form—is sodium chloride (NaCl). Sodium hydroxide (NaOH) is a common strong *base* and a common ingredient of commercial drain openers like Drano®. Baking powder and baking soda are both sodium bicarbonate ($NaHCO_3$), for which another name is *bicarbonate of soda*. Washing soda is sodium carbonate (Na_2CO_3) and can be added to laundry as a cleaning agent. Sodium carbonate is also called *soda ash*. Sodium peroxide is Na_2O_2 and is similar to hydrogen peroxide (H_2O_2) that is sold in pharmacies, except that sodium peroxide is a solid and hydrogen peroxide is sold in solution (usually at a concentration of 3 percent).

Sodium forms a number of phosphate compounds. Phosphoric *acid* (H_3PO_4) reacts with NaOH to form sodium dihydrogen phosphate (NaH_2PO_4) and water. Sodium dihydrogen phosphate in turn reacts with NaOH to form sodium monohydrogen phosphate (Na_2HPO_4) and water. Finally, sodium monohydrogen phosphate reacts with NaOH to form sodium phosphate (Na_3PO_4). These reactions are shown by the following equations:

$$H_3PO_4 \text{ (aq)} + NaOH \text{ (aq)} \rightarrow NaH_2PO_4 \text{ (aq)} + H_2O \text{ (l)}$$

$$NaH_2PO_4 \text{ (aq)} + NaOH \text{ (aq)} \rightarrow Na_2HPO_4 \text{ (aq)} + H_2O \text{ (l)}$$

$$Na_2HPO_4 \text{ (aq)} + NaOH \text{ (aq)} \rightarrow Na_3PO_4 \text{ (aq)} + H_2O \text{ (l)}.$$

Packaged as trisodium phosphate, or TSP, Na_3PO_4 is sold in hardware and paint stores as an abrasive cleaner in applications such as preparing previously painted surfaces for new paint jobs.

Sodium cyanide (NaCN) is poisonous (like all cyanides) and is used in the steel industry. Sodium nitrate ($NaNO_3$) is found in deposits of Chilean *saltpeter* and is used in fertilizers and explosives. Sodium sulfate can be an *anhydrous* substance (as Na_2SO_4) or a *hydrate* (as in $Na_2SO_4 \times 10H_2O$). In the latter case it is known as *Glauber's salt*.

In film photography, sodium thiosulfate ($Na_2S_2O_3$)—also known as sodium hyposulfite—is used to stop the developing process. Photo processors refer to it as *hypo.*

Sodium azide (NaN_3) is used in air bags in automobiles. Upon impact, an electrical discharge causes the sodium azide to rapidly decompose into sodium metal and nitrogen gas, as shown by the following equation:

$$2 \, NaN_3 \text{ (s)} \rightarrow 2 \, Na \text{ (s)} + 3 \, N_2 \text{ (g)}.$$

The nitrogen gas fills the air bag in a matter of about 40 milliseconds. The bag actually begins to deflate again before a passenger's head or torso strikes the bag. If the bag did not deflate again slightly, the inflated bag would be so hard as to injure the passenger. A slightly deflated bag, *(continued on page 22)*

THE PHYSICS OF SODIUM VAPOR LAMPS

Many street lights and building security lights give off a constant yellow glow. The light comes from electron energy level jumps in the sodium atoms, causing them to emit photons of either 589.0-nm or 589.6-nm wavelength. It is common to average the two energies, in which case the sodium emission wavelength is cited as 589.3 nm.

In a lamp filled with sodium gas, a voltage applied to electrodes at either end of the tube causes electrons to detach from the anode and accelerate toward the cathode. The electrons collide with the sodium vapor atoms, providing the energy for electrons within the sodium atoms to jump to higher levels. The atoms' rapid decay back to their ground state, with the simultaneous emission of photons, is called D-line emission simply because of an early alphabetical designation for various solar spectral lines observed by the German physicist Joseph von Fraunhofer (1787–1826).

Millennia of evolution have adapted the human eye to be most sensitive in the peak output region of the Sun's spectrum— the yellow wavelengths. This makes sodium light a good choice for streetlights and security lights in buildings, but, unlike sun-

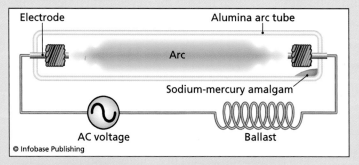

In a lamp filled with sodium gas, a voltage applied to electrodes at either end of the tube causes electrons to accelerate across the tube, exciting the gas that then emits photons of light.

light, it is nearly monochromatic. The single yellow wavelength is reflected differently by different objects, which makes it difficult for the human eye to distinguish colors as it normally does. This problem led to the development of high-pressure sodium vapor lamps. High pressure means more particles (electrons) per unit area within the lamp. Such a high density of electrons can allow for more energy exchange among the internal atomic electrons and the external collisional electrons. This has the effect of broadening the photon emission so it covers a larger region of the spectrum. Another way to broaden the spectrum is to add another atomic species that has a different spectrum. Mercury inclusion has shown good results in the manufacture of sodium vapor lamps for areas where nighttime color identification is important.

Sodium vapor street lamps are still common in many cities and are easily identifiable by their yellow glow. *(Richard Treptow/Photo Researchers, Inc.)*

(continued from page 19)

however, provides the protective cushioning that is desired. Potassium nitrate (KNO_3) and silicon dioxide (SiO_2) are also in the bag to quickly convert the sodium metal that is formed in the reaction to harmless sodium compounds. Otherwise, there would be the danger of metallic sodium exploding (which happens if it comes into contact with water).

The most common qualitative analytical technique to detect the presence of sodium in aqueous solutions or in solids (which can be dissolved in water) is a flame test. A wire made of an *inert* material like platinum or *nichrome* is dipped in a solution and then held in a flame. Sodium ions emit an intense yellow flame, readily verifying the presence of sodium in the sample.

SODIUM AND HEALTH

Sodium is crucial for maintaining proper fluid balances in humans and animals. Sodium ions are responsible for signal transmission across nerve-cell membranes and in regulating heart activity and other metabolic functions. It is not as important in plant systems, however, and is even toxic to many plant species. This is the reason that a vegetarian diet can lower sodium levels in the blood, and explains why deer, horses, and other livestock need natural or synthesized salt licks for health.

The sodium-to-water ratio in the body must be properly balanced or unfortunate effects may result. The *hypothalamus* area of the brain is fairly efficient at regulation of this ratio; when sodium is high, thirst results; when excess urination causes loss of water, the body also excretes sodium to help maintain the balance. People traveling long distances in desert or other dry climates may ingest salt ($NaCl$) as a way to retain water in their systems. The kidneys also play a role in sodium regulation.

These internal regulators can become overstressed, however, and ultimately unable to perfectly control the necessary balance. If sodium accumulates in the blood, too much water may be retained, which increases blood pressure in the arteries—a condition termed *hypertension*. *Arterioles* normally contract and expand to regulate blood flow, but excess Na constricts these blood vessels, and less blood can be returned to the heart. Such a physiological situation can lead to heart disease and stroke. To prevent hypertension from leading to more severe consequences, doctors prescribe high blood pressure medicines, often *diuretics,* to help flush excess sodium out of the system.

Modern eating habits, unfortunately, exacerbate the problem. Most processed foods incorporate high sodium content either to help preserve foods like meats and vegetables or to satisfy humans' increasing desire for a salty taste, as in potato chips or similar snacks. The U.S. Food and Drug Administration (FDA) recommends limiting daily sodium intake to around 2,300 mg, though medical practitioners would prefer it to be lower. In 2007, the American Medical Association (AMA) published a letter urging the FDA to reconsider its rating of salt as a food generally considered to be safe. There is argument from food manufacturers, however, and the debate continues. In March 2009 the Centers for Disease Control (CDC) issued a report stating that the majority of Americans should limit their salt intake to 1,500 mg per day—a quantity less than half the estimated consumer average.

TECHNOLOGY AND CURRENT USES OF SODIUM

Sodium is the most abundant alkali metal, and its most common salt (NaCl) has been used since prehistoric times to flavor and preserve

According to the American Medical Association, modern eating habits in the United States often incorporate too much salt. *(Shutterstock)*

SODIUM CONTENT OF SELECTED FOODS

DESCRIPTION	COMMON MEASURE	CONTENT PER MEASURE (MG)
alcoholic beverage, beer, light	12 fl oz	14
alcoholic beverage, beer, regular, all	12 fl oz	14
beans, baked, canned, plain or vegetarian	1 cup	856
beef stew, canned entrée	1 cup	947
biscuits, plain or buttermilk, prepared from recipe	4" biscuit	586
breakfast items, french toast with butter	2 slices	513
carrots, raw	1 carrot	50
cereals, ready-to-eat, Kellogg, Kellogg's Raisin Bran	1 cup	362
chickpeas (garbanzo beans, bengal gram), mature seeds, canned	1 cup	718
chickpeas (garbanzo beans, bengal gram), mature seeds, cooked, boiled, without salt	1 cup	11
danish pastry, cheese	1 danish	320
English muffins, plain, toasted, enriched, with calcium propionate (includes sourdough)	1 muffin	248
entrées, fish fillet, battered or breaded, and fried	1 fillet	484
fish, trout, rainbow, farmed, cooked, dry heat	3 oz.	36
fish, tuna salad	1 cup	824
grapes, red or green (European-type varieties, such as Thompson seedless), raw	1 cup	3
gravy, beef, canned	1/4 cup	326
macaroni and cheese, canned entrée	1 cup	1,061
pickles, cucumber, dill	1 pickle	833
pie crust, cookie-type, prepared from recipe, graham cracker, baked	1 pie shell	1,365

DESCRIPTION	COMMON MEASURE	CONTENT PER MEASURE (MG)
pork, cured ham, whole, separable lean only, roasted	3 oz	1,128
potatoes, mashed, home-prepared, whole milk and margarine added	1 cup	699
sandwiches and burgers, cheeseburger, large, single meat patty, with bacon and condiments	1 sandwich	1,043
soup, chicken vegetable, canned, chunky, ready-to-serve	1 cup	1,068

Source: USDA

food. Its ability to flavor food will always make sodium the member of the family in greatest demand. Many of the applications of sodium and its compounds could be fulfilled with the heavier alkali metals; sodium's abundance, however, makes it the least expensive. Although metallic sodium reacts vigorously with water, it is nevertheless much less hazardous to work with than the heavier alkalis.

In the form of the Na^+ ion, sodium is found in a wide variety of food products. Important sodium salts include the chloride, carbonate, bicarbonate, citrate, hypochlorite, chlorate, fluoride, nitrate, phosphate, sulfate, thiosulfate, hydroxide, and azide. Sodium chloride (NaCl, also known as *table salt* or *rock salt*) is used to cure fish, pack meat, cure hides, preserve canned foods, flavor foods, and melt ice on city roads. Sodium carbonate (Na_2CO_3, also called *soda ash*) is the ingredient in washing soda, which is used as a water softener for laundry. Sodium carbonate is also used in the glass industry and in the manufacture of soap, detergents, and other cleansers. Sodium bicarbonate ($NaHCO_3$) is a source of carbon dioxide and is the principal ingredient in baking powder and baking soda. Sodium citrate ($Na_3C_6H_5O_7$) is added to soft drinks to control their acidity.

Sodium hypochlorite (NaClO) is the source of chlorine in household bleach. Sodium chlorate ($NaClO_3$) is used in weed killers, matches, and explosives. Sodium fluoride (NaF) is added to toothpaste to provide protection against dental cavities. Sodium nitrate ($NaNO_3$, also known as Chile *saltpeter*) is an important source of nitrogen in fertilizers. Trisodium phosphate (Na_3PO_4, sold as "TSP") is used in cleaning and in laundry as a water softener. It is an excellent degreaser and frequently used to prepare surfaces for painting. Sodium sulfate (Na_2SO_4, also known as *Glauber's salt*) is used in the manufacture of glass, pulp, detergents, ceramics, and pharmaceuticals. Sodium thiosulfate ($Na_2S_2O_3$, also known as *hypo*) is used as the *fixing agent* in film photography. Sodium hydroxide (NaOH, also called *caustic soda*) is a strong base and the principal component of lye and some brands of drain openers. It is also an important compound in the synthesis of industrial chemicals and in the manufacture of soap, pulp, and paper. Sodium azide (NaN_3) is used as the source of nitrogen gas to inflate automobile air bags.

Excitation of sodium atoms produces the yellow light seen in fireworks and street lamps. The subject of sodium atoms is found in much current scientific research. The transfer of angular momentum from laser light to particles was first accomplished in a small cloud of sodium atoms.

Sodium metal has a very high reactivity (although it is not as reactive or as hazardous as potassium, rubidium, or cesium). An important use of the metal is to reduce the ores of less *active metals* to their neutral state. Examples include zinc, chromium, and titanium.

Other uses of sodium include the following examples. Sodium is used to reduce animal and vegetable oils to compounds that serve in the manufacture of detergents. Sodium is used to manufacture sodium hydride (NaH), which is used for syntheses of organic compounds and sodium cyanide (NaCN), the latter of which is used by miners to extract gold from its ores and by entomologists to kill insect specimens without damaging them. Liquid metallic sodium acts as a heat-transfer (cooling) agent in fast-neutron nuclear reactors, where water cannot be used because water would slow down the neutrons.

Regardless of sodium's applications in industry, the most important fact about sodium is that it is an element essential to health in humans and other higher animals. The sodium ion (Na^+) is present in greatest concentration in the fluids that are outside the cells of animals (potassium ions are present in greater concentration inside cells). Together, Na$^+$ and K$^+$ regulate conduction of nerve impulses, *osmotic pressure* inside cells, and transport of ions across membranes. Because sodium ions are so easily excreted from the body in sweat, in urine, and in feces, diet must supply a daily replenishment of sodium levels. Otherwise, severe depletion of sodium can result in weakness, muscle cramps, nausea, and decreased blood pressure.

Clearly, salt production is a major industry. Worldwide, roughly 220 million tons (200 million metric tons) of sodium salt are recovered yearly from seawater, rock salt, and inland salt waters. With 60 percent being used by the chemical industry, 30 percent by the food industry, and 10 percent by other end users (such as municipalities to deice roads), salt will continue to play an important role in human society.

3

Potassium

Potassium, element number 19, has a density of 0.86 g/cm³ and is so soft that it can be cut with a butter knife. It is even more reactive with water and oxygen than lithium or sodium are, so that potassium also is never found as the pure metal. In addition, for safety reasons, it is rarely handled as the pure metal. In part for safety reasons, there is little demand in industry for potassium metal itself except to make sodium-potassium alloys.

The main source of potassium is potash rock, a group of *sedimentary* substances often found in association with rock salt, gypsum, and dolomite. The term *potash* itself was derived originally from the pot ash formed when seawater was mixed with wood ashes and boiled. Because of the importance of potassium for plant growth, about 95 percent of the world's potash is converted into fertilizer.

THE BASICS OF POTASSIUM

	Potassium	
2		63.38°
8	K_{19}	759°
8		
1		
	+1	
	39.0983	
	0.0000123%	

Symbol: K
Atomic number: 19
Atomic mass: 39.0983
Electronic configuration: [Ar]4s^1

T_{melt} = 145°F (63°C)
T_{boil} = 1,398°F (759°C)

Abundances:
In Earth's crust 18,400 ppm
In seawater 380 ppm

Isotope	Z	N	Relative Abundance
$^{39}_{19}$K	19	20	93.2581%
$^{40}_{19}$K	19	21	0.0117%
$^{41}_{19}$K	19	22	6.7302%

In this chapter, the reader will learn about the synthesis of potassium in stars, the geochemistry of potassium and its role in dating rock samples, the chemistry of potassium, its role in nutrition, and important uses of potassium in society.

THE ASTROPHYSICS OF POTASSIUM

Almost all experimental data show the abundance of potassium in meteorites, stars of the galactic disk, and dust in the interstellar medium to be nearly the same as the solar abundance. Such a ubiquitous abundance of an element is relatively rare and is interpreted by astrophysicists as a sign that explosive oxygen burning must play an important role in the production of potassium. Explosive oxygen burning can only take place in the violent explosions of Type II supernovae. This means that within the brief explosive interval the atmosphere of the star becomes hot enough for oxygen atoms to fuse with each other and then react with ambient protons and neutrons to form potassium nuclei, which are then distributed into the cosmos by the force of the blast.

A certain class of pseudostars known as "brown dwarfs," however, is known for an excess of potassium in their outer atmospheres. Brown dwarfs are gaseous objects about the diameter of Jupiter (only much more massive) that didn't quite collect enough gravitational mass to maintain the fusion temperature for hydrogen. They typically glow in the red wavelengths from the heat of gravitational contraction, though the less massive "T-dwarfs" glow purple-red because of the potassium atoms that absorb most of the light in the green portion of the spectrum. More data need to be acquired and examined to understand the potassium content of these stellar atmospheres.

The surfaces of two moons of Jupiter, Europa and Io, are also of current interest, partly due to unexpected potassium abundance and partly due to the proximity to Earth. Europa's atmospheric potassium content is intriguing because its origin is in doubt. For some time, astronomers have suggested that the potassium (and sodium) content in its atmosphere must be a result of its gravitational acquisition of those trace elements from the atmosphere of the Jovian moon, Io. New results, however, indicate that Europa's ratio of atmospheric Na/K is at least twice as high as Io's. Michael Brown, of the Division of Geological and Planetary Sciences at the California Institute of Technology, concludes "that the trace elements sputtered into Europa's extended atmosphere are intrinsic to the surface of Europa." Potassium-argon dating techniques may help understand this Jovian moon's history.

Fortunately for astronomers, Europa's surface has not regenerated recently. At an age of at least 10 million years, its surface can tell much about its history. NASA and others are showing interest in Europa's potassium as well as the possibility that life processes should be investigated. Europa may be hiding an ocean beneath its surface. Lander missions are being considered, which would be of immense usefulness in understanding one of the most dynamic bodies of our solar system.

POTASSIUM ON EARTH

Potassium is a component of many of Earth's minerals, including muriate of potash or sylvite (KCl), carnallite (KCl $\times$ MgCl$_2$ $\times$ 6H$_2$O), lang-

Potassium feldspar crystals are larger than the other minerals in this rock matrix located at Cathedral Peak in Yosemite National Park. *(USGS)*

beinite ($K_2Mg_2(SO_4)_5$), and polyhalite ($K_2Ca_2Mg(SO_4)_4 \times 2H_2O$). The term *potash* is used rather loosely, but usually refers to potassium carbonate (K_2CO_3), a fine white powder that is manufactured for use in making soap and glass. Caustic potash is the common name for potassium hydroxide (KOH), which has myriad uses and is mined mainly in the western United States and Germany.

At only 2 percent abundance, the potassium content of Earth's crust is much lower than the average measured in meteorites, most of which formed during the earliest stages of the solar system about 4.5 billion years ago. Since Earth formed at roughly the same time, the disparity means that some potassium-depleting process must have occurred as the planet evolved. Scientists have long speculated that much of the potassium may have moved into the molten outer core that surrounds the solid inner core of Earth. The molten core is composed primarily of liquid iron that flows as the planet rotates. This flow of iron ions produces the strong magnetic field that surrounds the Earth. The magnetic field, in turn, protects plants and animals from solar wind protons and cosmic rays by deflecting them away from the Earth's surface.

The iron would not be molten and flow, however, without a continuous heat source to keep it from solidifying. Radioactive fission energy from uranium and thorium decay has long been the assumed source of the heat in the core, and still provides that heat, but not enough to have sustained a magnetic field for 4 billion years.

Potassium 40, with a half-life of 1.25 billion years, could provide the added energy needed, but potassium will not normally mix with iron. Experiments at the Lawrence Berkeley and Stanford Synchrotron Radiation Laboratories, however, have shown that at extreme temperatures and pressures like those that exist ca. 373–435 miles (600–700 km) beneath Earth's surface, a different situation may be possible. With temperatures around 4,500°F (2,480°C) and pressures near 4 million pounds per square inch (2.8 million newtons/cm^2), a potassium atom's outer electron can be pressed in so close to the next lower electron shell that it can behave more like an iron atom. Potassium 40 thus squeezed and mixed into molten iron could have increased the *flux* of heat from the mantle to the core by up to 20 percent—enough to provide the energy required to sustain Earth's magnetic field over its history. Such dynamic mixing during the planet's development would also explain the leaching and depletion of potassium from Earth's crust. According to a 2003 U.C. Berkeley press release, Mark Bukowinski (who predicted the K/Fe alloy more than three decades ago) said these experiments have "demonstrated that potassium may be an important heat source for the geodynamo, provided a way out of some troublesome aspects of the core's thermal evolution, and further demonstrated that modern computational mineral physics not only complements experimental work, but that it can provide guidance to fruitful experimental explorations."

THE CHEMISTRY OF POTASSIUM

Potassium forms virtually all the same chemical compounds that sodium does. Common potassium compounds are potassium chloride (KCl), potassium bromide (KBr), potassium iodide (KI), potassium carbonate

(K$_2$CO$_3$), potassium nitrate (KNO$_3$), potassium sulfate (K$_2$SO$_4$), potassium carbonate (K$_2$CO$_3$), and potassium hydroxide (KOH), which is a strong caustic base like sodium hydroxide. All of these compounds are soluble in water. As a general rule, the potassium ion itself is unreactive in aqueous solutions.

The natural form of potassium nitrate—called *niter* or *saltpeter*—is one of the three components of gunpowder (today commonly referred to as "black powder"), which consists of 75 percent KNO$_3$, 15 percent charcoal, and 10 percent sulfur. The three components are ground together into a very fine powder.

Potassium, as K$^+$, is an essential element for both humans and other animals that plays an important role in metabolic processes, especially in its ability to accelerate chemical reactions in muscles and provide muscles with their energy. Potassium and sodium together

Avocadoes, potatoes, and bananas all have a high potassium content. *(Tobi Zausner, Ph.D.)*

POTASSIUM-ARGON DATING

For dating archaeological objects, such as animal bones, the most commonly known technique is carbon dating. This method, however, is useful only for dating objects younger than about 40,000 years that were once living or can be logically connected to previously living organisms.

Scientists have been able to trace the origin of the human species back nearly 2 million years by another method—K/Ar dating. Unlike carbon dating, which relies on measuring how much carbon 14 has decayed, the K/Ar method counts the number of argon 40 atoms produced in the following reaction, which occurs in 11 percent of ^{40}K decays:

$$^{40}_{19}K \rightarrow {}^{40}_{18}Ar + e^-.$$

Since argon is inert, it remains intact until it is released, which is usually accomplished by heating the object of interest. There are important limitations to this method, so it can only be used under specific circumstances. Clearly, the scheme is viable only for samples that originally contained a considerable fraction of potassium 40. This is most common in volcanic rock and ash, as lava emerging from Earth's molten core carries a higher

help regulate the flow of water in the body. A diet that includes fruit, vegetables, and meat should provide all the potassium a healthy human requires.

The most common qualitative analytical test for the presence of potassium ions in an unknown sample is a flame test. If potassium is present, a wire that has been dipped in a solution will give a violet flame. Because sodium is likely also to be present, the yellow sodium flame will mask potassium's violet flame, since sodium's color is more intense. Therefore, the flame is viewed through a special glass called "cobalt-blue glass" that absorbs sodium's yellow color while transmitting potassium's violet color.

abundance of potassium than Earth's crust. Because heat frees any argon gas, the date revealed by counting argon atoms will tell how long ago the lava cooled. This aspect is also important in that it can tell the era of the last use of cooking implements discovered at ancient settlements.

Potassium-argon dating cannot give a definitive age for fossils, bones, or petrified trees, as these do not contain sufficient potassium 40. A clever evaluation of geological layers, however, can provide a range of dates for human artifacts. For many sites, a nearby volcano may have erupted several times throughout its history. In such regions an archaeologist may find remains of human settlements embedded between layers of volcanic rock, where the lower is older than the upper layer. A famous example is the Olduvai Gorge site in Tanzania, where the 1968 discovery of *Homo habilis* tested the worth of the (then controversial) potassium-argon dating. The method dated the human fossils at 1.75 million years, the currently accepted age of humanity's ancient cousins. Potassium-argon dating is now considered the most useful technique for finding the age of potassium-containing deposits between 1 million and 5 million years old.

POTASSIUM AND HEALTH

Potassium in its ionic form, K^+, is the most abundant positive ion in human and animal cells. As an electrolytic solution, K^+ ions are pumped through the blood to all vital organs. Potassium's importance to the physiological system cannot be overstated: It plays a crucial role in electrical pulse transmission along nerve fibers; protein synthesis; acid-base balance; and formation of collagen, elastin, and muscle.

Potassium is highly soluble in water and regulates flow across semipermeable membranes like cell walls. It is this feature that makes a deficiency or an excess of potassium hazardous to health. Either extreme can have undesirable and even disastrous consequences.

A diet with sufficient potassium can reduce the risk of stroke, slow the progress of kidney disease, and control blood pressure. Americans consume on average 30–40 percent less potassium than has been the norm for the human race over the course of its history. The behavior is attributable to the lowered consumption of fresh fruits and vegetables—a consequence of the easy availability of processed foods. The new trends in diet have produced a potassium deficiency in large segments of the population, which has in turn led to an increased incidence of hypertension, cardiovascular disease, kidney failure, diabetes, arthritis, and osteoporosis. The use of diuretics, laxatives, and steroids also contributes to a deficiency, as too much potassium can be flushed out of the body.

Supplementation in pill form, however, is not the solution. Introducing concentrated potassium into the body can have devastating consequences. Potassium content in the blood regulates flow across cell

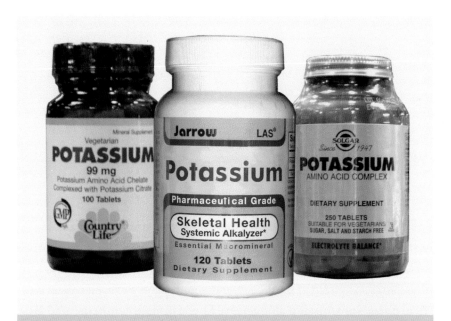

Excess potassium supplementation can be risky. The consensus among medical practitioners is that the best way to get the right amount of potassium is by eating noncereal fruits and vegetables. *(Tobi Zausner, Ph.D.)*

POTASSIUM CONTENT OF VARIOUS FOODS

FOOD	SERVING SIZE	POTASSIUM (MG)
almond	2 ounces	412
apricots, dried	10 halves	407
artichoke	1 cup	595
avocados, raw	1 ounce	180
bananas, raw	1 cup	594
beans, baked	1 cup	752
beans, kidney	1 cup	713
beans, lima	1 cup	955
beans, pinto	1 cup	800
beets, cooked	1 cup	519
black-eyed peas	1 cup	690
brussel sprouts, cooked	1 cup	504
cantaloupe	1 cup	494
chick peas	1 cup	477
dates, dry	5 dates	271
figs, dry	2 figs	271
kiwi fruit, raw	1 medium	252
lentils	1 cup	731
melons, honeydew	1 cup	461
milk, fat free or skim	1 cup	407
orange	1 orange	237
pears (fresh)	1 pear	208
peanuts, dry roasted and unsalted	2 ounces	374
potato, baked	1 potato	1,081
prunes, dried	1 cup	828
raisins	1 cup	1,089
spinach, cooked	1 cup	839
winter squash	1 cup	896

Source: USDA Nutrient Database for Standard Reference, Release 15. Available online. URL: http://www.nal.usda.gov/fnic/foodcomp/Data/SR15/sr15.html. Accessed on August 13, 2009.

walls, which regulates nerve signals. An excess of potassium can lead to failure of the system to transmit signals to vital organs, and result in death. In particular, patients on dialysis for kidney failure must be very cautious in their potassium consumption. The consensus among medical practitioners is that the best way to get the right amount of potassium is by eating noncereal fruits and vegetables, with 4.7 grams per day recommended as a minimum for adults.

TECHNOLOGY AND CURRENT USES OF POTASSIUM

Potassium metal reacts more violently with water and with air than sodium does, making potassium more dangerous to handle. Because it is safer and less expensive to use sodium in situations where either metal would suffice, potassium metal itself is seldom used. (Elementary sodium production exceeds potassium production by a factor of 1,000.) Potassium metal's main use is to manufacture potassium superoxide (KO_2), which is formed by chemical reaction between potassium and atmospheric oxygen (O_2). (The superoxide ion [O_2^-] is an ion derived from O_2 that is common in biological systems, but uncommon apart from living organisms.) Potassium superoxide can be used to generate oxygen gas, as shown in the following reaction:

$$4\ KO_2\ (s) + 2H_2O\ (l) \rightarrow 4\ KOH\ (aq) + 3\ O_2.$$

KO_2 is used in equipment for patients requiring supplemental oxygen, in mine rescue equipment, and in the space program. The Russian Space Agency has successfully used KO_2 in spacesuits and in the Soyuz spacecraft.

A number of compounds contain the K^+ ion, and in many cases have the same uses as the corresponding sodium compounds. Examples in which K^+ is found include potassium chloride, bromide, iodide, cyanate, nitrate, sulfate, acetate, carbonate, chlorate, chromate, dichromate, ferricyanide, cyanide, hydroxide, perchlorate, permanganate, and persulfate. In household and kitchen products, potassium chloride (KCl) substitutes for NaCl for people with sensitivity to the Na^+ in ordinary table salt. As a replacement for sodium chloride, potassium chloride may be recommended to reduce high blood pressure or to reduce water retention. Potassium bromide (KBr) is a sedative. Potassium iodide (KI) is added to table salt to make "iodized salt." It may also be used as a source of iodine in disinfectants. In addition, KI stimulates saliva production and loosens and breaks up nasal mucous congestion.

Some potassium compounds have applications in gardening and agriculture. Potassium cyanate (KOCN) is applied to lawns to kill crab-grass. Nearly all (95 percent) of potassium mined goes into fertilizers. Important sources of potassium for fertilizers include potassium chloride, potassium nitrate (KNO_3, also known as *saltpeter* or *niter*), and potassium sulfate (K_2SO_4); in KNO_3, nitrate (NO_3^-) also supplies plants with nitrogen. Potassium nitrate is also a component of match heads, gunpowder, and explosives.

Potassium acetate ($KC_2H_3O_2$) is produced in large quantities as a raw material for manufacturing penicillin. Potassium carbonate (K_2CO_3) is used in glassmaking. Potassium chromate (K_2CrO_4) is a yellow substance used to dye textiles and to add color to ink. Potassium dichromate ($K_2Cr_2O_7$) is an orange substance used to color glass, preserve wood, and develop blueprints. Film developing and printing processes may use potassium ferricyanide [($K_3Fe(CN)_6$), a bright red compound] to remove silver from negatives and to enhance color.

Potassium cyanide (KCN) is extremely toxic and is used as a pesticide and as a *fumigant*. Mystery readers and movie fans may recognize cyanide's poisonous role in murder and suicide. When KCN becomes moist (as occurs when it comes into contact with saliva), it converts to hydrogen cyanide (HCN), which inhibits respiration and results in convulsions and heart failure. Hydrogen cyanide is a gas with an odor that resembles the odor of almond pits (because they contain tiny amounts of cyanide). A seemingly more benign use of KCN is to extract gold and silver from their ores, although in that case the cyanide becomes an environmental concern.

Potassium hydroxide (KOH) is used to manufacture soaps, drugs, alkaline batteries, adhesives, and fertilizers. Like sodium hydroxide, KOH can be used in drain openers. Potassium perchlorate ($KClO_4$) is used in fireworks, sparklers, explosives, and solid rocket propellants.

Potassium permanganate (KMnO$_4$) is a purple substance that is used as a disinfectant and to kill bacteria. Potassium persulfate (K$_2$S$_2$O$_8$) is a bleaching agent and an *antiseptic*.

About 33 million tons (30 million metric tons) of potassium (measured as K$_2$O) are produced worldwide annually, mostly in Canada, the United States, and Chile. As an important component of fertilizer and an essential nutrient in the human diet, the demand for potassium products may be expected to continue.

4

Rubidium, Cesium, and Francium

Rubidium, cesium (spelled "caesium" in England), and francium are soft, silvery gray metals. The abundances of these three elements are far less than the abundances of lithium, sodium, and potassium. The following is a list of three notable characteristics of these elements:

- rubidium is element number 37 with a density of 1.53 g/cm³, making it denser than water (unlike lithium, sodium, and potassium, which are less dense than water);
- cesium is element 55 and is about 30 times less abundant than rubidium, with a density of 1.88 g/cm³;
- francium is element 87. Because all of the isotopes of francium are radioactive with very short half-lives, francium is extremely scarce, and little study has been made of its properties. In fact, so little francium has ever existed at one time that its density has never been measured.

THE BASICS OF RUBIDIUM

Symbol: Rb
Atomic number: 37
Atomic mass: 85.4678
Electronic configuration: $[Kr]5s^1$

T_{melt} = 102°F (39°C)
T_{boil} = 1,270°F (688°C)

Abundances:
In Earth's crust 78 ppm
In seawater 0.12 ppm

Rubidium		
2		39.31°
8	**Rb**$_{37}$	688°
18		
8		
1		
	+1	
	85.4678	
	2.31×10^{-8}%	

Isotope	Z	N	Relative Abundance
$^{85}_{37}$Rb	37	48	72.17%
$^{87}_{37}$Rb	37	50	27.83%

THE BASICS OF CESIUM

Symbol: Cs
Atomic number: 55
Atomic mass: 132.90545
Electronic configuration: $[Xe]6s^1$

T_{melt} = 82°F (28°C)
T_{boil} = 1,240°F (671°C)

Abundances:
In Earth's crust 2.6 ppm
In seawater 0.0005 ppm

Cesium		
2		28.44°
8	**Cs**$_{55}$	671°
18		
18		
8		
1	+1	
	132.90545	
	1.21×10^{-9}%	

Isotope	Z	N	Relative Abundance
$^{133}_{55}$Cs	55	78	100%

THE BASICS OF FRANCIUM

	Francium	
2		27°
8		
18	**Fr**$_{87}$	
32		
18		
8	+1	
1	[223]	

Symbol: Fr
Atomic number: 87
Atomic mass: No stable isotopes.
(The only isotope found in Earth's crust is ^{223}Fr.)
Electronic configuration: [Rn]7s^1

T_{melt} = 81°F (27°C)
T_{boil} = unknown

Abundances:
In Earth's crust trace
In seawater negligible

Isotope	Z	N	Relative Abundance
$^{223}_{87}$Fr	87	136	100%

Note: all isotopes of francium are radioactive with very short half-lives; $^{223}_{87}$Fr is the most common isotope; it has a half-life of 21 minutes.

The reactivity of alkali metals increases upon descending the column. By the time rubidium and cesium are reached, contact with the moisture in air would be explosive.

In this chapter the reader will learn about the syntheses of rubidium, cesium, and francium in stars, the chemistry of rubidium and cesium, the use of cesium in atomic clocks, and other applications of rubidium and cesium. Too little francium exists for it to have any practical uses.

THE ASTROPHYSICS OF HEAVY ALKALI METALS

While francium is produced only in supernovae, both rubidium and cesium are synthesized when the nuclei of iron and heavier elements within stars capture slow-moving free neutrons. This is called the "s-process." The free neutrons are released in the following alpha-particle capture reactions in some stellar interiors.

Cesium is a soft, silvery gray metal.
(Lester V. Bergman/CORBIS)

$$^{13}_{6}\text{C} + {}^{4}_{2}\alpha \rightarrow {}^{16}_{8}\text{O} + {}^{1}_{0}\text{n}$$

$$^{22}_{10}\text{Ne} + {}^{4}_{2}\alpha \rightarrow {}^{25}_{12}\text{Mg} + {}^{1}_{0}\text{n}$$

Which reaction dominates depends on the ambient temperature in the interaction region: Below about 100 million K, the carbon process dominates, whereas above that temperature the neon process tends to produce the slow neutrons. Since temperatures are typically higher in stars with more mass—as higher gravitational force accelerates and heats the stellar material—the neon reaction is more likely to proceed in stars much heavier than the Sun. The temperature of a distant star cannot, however, be measured directly. Rubidium production happens to be sensitive to neutron density—which is determined by the neutron production process—and can therefore be a good indicator of which reaction prevails in particular stars, especially when viewed as a ratio with other available s-process elements like yttrium or strontium.

An unexpectedly high abundance of rubidium has been observed in *asymptotic giant branch* (AGB) stars with masses between about four and eight times that of the Sun. These stars are unusually active and exhibit thermal pulses that make for a high neutron density. Formation of the radioactive isotope ^{87}Rb is, therefore, more likely than in cooler stars. Rubidium 87 has a half-life of 47 billion years, so once formed will remain with the star until the end of its life. Unfortunately, it has not been possible to measure the ^{87}Rb content in stars.

Measurements of the ^{85}Rb/^{87}Rb ratio in the interstellar medium have been achieved but give puzzling results. The interstellar value turns out to be about half that measured in meteoritic samples. This is an important discrepancy because radioactive dating methods for meteorites assume that there was no exchange of rubidium during the time when the meteorites formed in the early solar system. Scientists are now considering the possibility that the rubidium isotope fraction may have been affected by nearby AGB stars.

By contrast, in the disk and halo of the Milky Way, some metal-poor stars that are also overabundant in rubidium relative to iron are confusingly deficient in only slightly heavier elements, such as yttrium and zirconium. This would seem to indicate that, in these particular stars, the rubidium production occurs via the r-process (see chapter 8), which is not capable of producing the heavier elements.

Overall, much work remains to be done before the astrophysical abundances of the heavy alkali elements are properly understood.

THE CHEMISTRY OF RUBIDIUM AND CESIUM

Because rubidium and cesium are extremely active metals, they are never found in nature as the pure elements. Rubidium is so extremely reactive with oxygen that samples of the metal will ignite spontaneously in an atmosphere of pure oxygen. When melted, the molten metal will ignite spontaneously with oxygen at its 21 percent concentration in air. Rubidium will even react explosively with ice at temperatures as low as −150°F (−101°C). Rubidium reacts so strongly with bromine or chlorine that a flame can be observed. For obvious reasons, rubidium is such a hazardous substance that people handle it only when following strict safety procedures observed by industry and government laboratories.

Cesium is a soft metal with a low melting point. It is the most chemically reactive metal and, even more strongly than rubidium, combines readily with oxygen, with the halogens, and with water. Cesium will react with ice to a temperature as low as −177°F (−116°C).

In compounds, rubidium and cesium exist only as the +1 ions Rb^+ and Cs^+. Representative compounds include the following:

Chloride	Nitrate	Oxide	Carbonate	Sulfate
RbCl	$RbNO_3$	Rb_2O	Rb_2CO_3	Rb_2SO_4
CsCl	$CsNO_3$	Cs_2O	Cs_2CO_3	Cs_2SO_4

All of these compounds are soluble in water.

CESIUM AND ATOMIC CLOCKS

Since the beginning of timekeeping, the clock has been based on a measurement of periodic motion. For centuries, humans marked time by the rotation of the Earth. Sunrise and sunset determined when it was time to hunt, gather, or farm. The solar day is variable because of Earth's tilt on its axis and cannot provide the accurate measure of hours, minutes, and seconds needed by a more advanced society that is able to light the evenings by its own means.

Pendulums were more accurate, followed by quartz oscillators. In 1880, Pierre and Jacques Curie discovered that sending a current through a quartz crystal could result in a resonance situation with cyclic behavior, making a quartz oscillator that could be used to mark time. Because crystals grow in miniature, quartz crystal watches became the standard in the 1960s. But the periodic nature of resonant quartz excitations depends on the shape of each crystal as well as the ambient temperature and humidity. Every crystal behaves differently, and none can constitute a reputable standard for the accuracy needed by global positioning systems, for example.

Atoms, however, display an extremely precise distinction in energy levels. Since energy is related to frequency f by the following equation,

$$f = \frac{E_2 - E_1}{h},$$

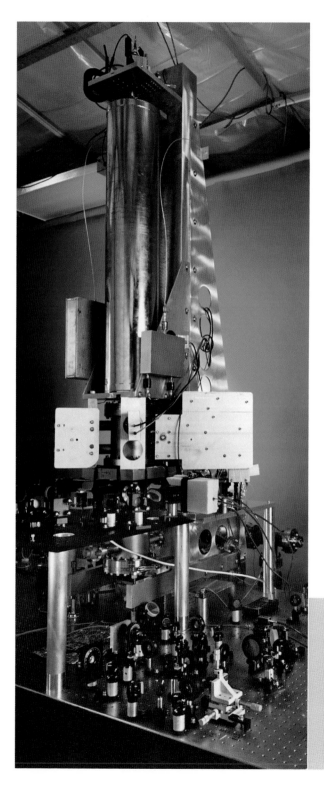

NIST-F1, the nation's primary time and frequency standard, is a cesium fountain atomic clock developed at the NIST laboratories in Boulder, Colorado. *(Geoffrey Wheeler Photography/NIST)*

THE HUMAN BODY: NO PLACE FOR THESE ELEMENTS

Rubidium typically exists in the human body at the level of only 1/1,000 of 1 percent, and cesium content is even lower. Rubidium and cesium are both absorbed from soil by plants and are, therefore, present in small quantities in vegetables and up the food chain to meat products and humans. Rubidium is known to stimulate mammalian metabolism, probably because of its physical and chemical similarity to potassium, which plays a crucial role in electrical pulse transmission along nerve fibers; protein synthesis; acid-base balance; and formation of collagen, elastin, and muscle. Its likeness to potassium may be the reason rubidium enhances growth in some plants. For particular insects, however, the introduction in the laboratory of rubidium to the bloodstream has been shown to drastically reduce fluid secretion and to change the electric potential across cell membranes. Excess rubidium is almost never encountered, however, in nature.

Cesium, on the other hand, is toxic to plants in anything but trace amounts, whereas indications are that Cs^+ ions impair the activity of potassium-binding sites in proteins. Excess cesium can be found in the air and in soils as a by-product of nuclear testing and spent nuclear fuels. Radioactive cesium 137, which results from the fission of uranium 235, decays by emission of a

where the subscripts denote higher and lower energy levels and h is a constant, atomic excitations are an excellent resource for periodic measurements.

Tiny energy transitions in atoms take place because the spin of the nucleus creates minuscule magnetic fields. Though small, these fields are discernible to the atom's electrons, which means that every available energy level is actually a range of energies separated by differences in the *microwave* range, discernible only in high-resolution experiments.

fast electron (ionizing radiation) and has a half-life of 30 years. Because cesium is so similar to potassium, it is easily absorbed into the human or animal system and takes three to four months to be excreted. During that time, the body is at increased risk of developing cancer, as the emitted electrons may interact with and mutate normal, healthy cells.

Another isotope, however, is beneficial for treating cancer, especially of the prostate. Cesium 131 has a half-life of only 10 days and emits low-energy photons that can be used to destroy cancer cells. Injection of a small source or "seed" of the *radioiso-tope* to a specific site—a process called *brachytherapy*—allows radiation to be aimed like bullets that can damage target recep-tor cells. It is especially potent against those that divide rapidly like cancer cells.

The radioactive isotope rubidium 82—another photon emitter—is also finding a useful niche in diagnostic medicine. As the tracer in *PET (positron emission tomography)* scans, rubidium is cheaper than the more commonly used ammonia, which must be produced in a particle accelerator. Combined with a *CT scan* that helps visualize blood flow, this method is currently the most accurate when imaging blood flow restrictions that may lead to cardiac arrest.

Microwave generation became available in the 1930s and 40s, and scientists quickly realized the importance for atomic excitation experiments. Like hydrogen, cesium has one easily excitable electron in its outer shell, but is heavier and therefore does not move as fast. A gas of cesium atoms will stay localized in an experimental chamber ready to receive microwave signals, and cesium just happens to have an easily attainable excitation energy. This energy is so precise that it can only be measured by a microwave energy scan of a chamber full of cesium 133 atoms. When the maximum number of excitations is detected, the most accurate excitation energy is realized, which turns out to be 9,192,631,770 cycles per second. To set the atomic clock, this frequency is fed into a quartz crystal oscillator to make a signal that appears at one pulse per second.

The process is still not perfect, as it can only tell time with an accuracy of 1 part in 10 trillion—within one second every 300,000 years. One of the impediments to higher accuracy is the natural random thermal motion of the atoms in the chamber. Collisions due to this motion lead to local density changes that are unpredictable. Cooling down the cesium gas by standard refrigeration methods is not enough. Physicists are now able to slow atoms by collisions with photons of very specific wavelength, however. Lasers aimed from six perpendicular directions can effectively slow and trap atoms in an experimental chamber— research for which Steven Chu, Claude Cohen-Tannoudji, and William Phillips were awarded the 1997 Nobel Prize in physics. Laser-cooling methods currently being tested have led to a new accuracy standard of 4 parts in 10^{16}. Such precision may lead to more rigorous tests of general relativity and help in understanding variations in pulsars.

RUBIDIUM AND LASER-COOLING

Rubidium gas has become important in the study of an exotic state of matter called a *Bose-Einstein condensate*. This state, first predicted in 1924 by Indian physicist Satyendra Nath Bose, was not observed until 1995. Many laboratories now produce these cooled clouds of atoms, mostly using gases of alkali elements, which have appropriate spin and magnetic properties.

Atoms sent into an experimental chamber can be slowed by collisions with photons of very specific wavelength by lasers aimed from six perpendicular directions. The laser energy must be tuned close to an excitation energy of the atomic species, and rubidium 87 happens to have an easily accessible transition at 780 nm. When the excited atomic electrons in rubidium jump back down to their ground state, they emit photons in the visible range, so it is easy to see the atomic cloud as it is slowed and trapped by an applied magnetic field.

Bose-Einstein condensates display unusual properties and are considered a new phase of matter. The atoms get packed together so closely that their wavefunctions become correlated like those of photons in

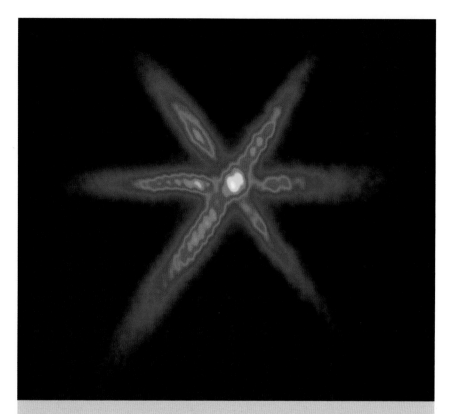

Low-light fluorescence image of 1 million trapped rubidium atoms cooled to micro Kelvin temperatures. The gaseous atoms are confined at the intersection of three focused, off-resonant laser beams, and the image was taken by briefly illuminating the atoms with resonant laser light. *(Andor Technology)*

a laser beam, and coherent matter waves can be formed. Though no practical uses of this state of matter have yet been developed, it is considered cutting-edge to work to understand the behavior, such as oscillation modes and thermodynamics of these systems. The 1997 Nobel Prize in physics was awarded to Steven Chu, Claude Cohen-Tannoudji, and William Phillips for their research in cooling and trapping atoms.

TECHNOLOGY AND CURRENT USES OF RUBIDIUM AND CESIUM

Rubidium and cesium react readily with oxygen and explosively with water, which makes them hazardous to work with in elemental form.

In addition, since most of their compounds would duplicate analogous sodium and potassium compounds, there are relatively few commercial uses of rubidium or cesium. In areas in which they do find application, generally speaking, cesium is preferred over rubidium. Cesium, and to a lesser extent rubidium, is used in photocells, televisions, motion picture equipment, and luminescent screens. Cesium is also used in spectrographic instruments and vapor arc lamps.

Rubidium and cesium are used in very accurate clocks. These clocks have been used to test phenomena predicted by *general relativity* theory, to investigate changes in the frequencies of *pulsars,* to track missiles and satellites, and to maintain the extremely accurate time-keeping required by *global positioning systems.*

A few salts of rubidium and cesium find application—for example the use of rubidium carbonate (Rb_2CO_3) and cesium carbonate (Cs_2CO_3) in the manufacture of glass and ceramics. Cesium fluoride (CsF) and cesium iodide (CsI) absorb X-rays and gamma rays and can be used in medical diagnostic equipment. In the form of cesium chloride (CsCl), the radioactive isotope cesium 137 is a source of gamma rays used in cancer treatment. In addition, cesium 137 is used in educational physics laboratory classes to study radioactivity.

At the present time, worldwide rubidium production is small. Currently more expensive than gold or platinum, rubidium finds little demand. Worldwide, only about 22 tons (20 metric tons) of cesium compounds are produced annually. Given the prevalence of sodium and potassium usage, production or use of either rubidium or cesium is not likely to increase.

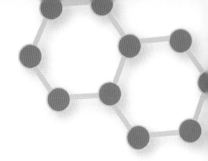

PART

II

ALKALINE EARTH METALS

INTRODUCTION TO ALKALINE EARTH METALS

The alkaline earth group as a whole stands in marked contrast to transition metals and post-transition metals. For example, most of the metals in the periodic table form insoluble precipitates with the sulfide ion (S^{2-}), with the result that sulfide ores of transition and post-transition metals are very common in Earth's crust. Common examples of metal sulfides include galena (lead sulfide), cinnabar (mercury sulfide), greenockite (cadmium sulfide), acanthite (silver sulfide), cobaltite (cobalt arsenic sulfide), sphalerite (ZnS), stibnite (antimony sulfide), several copper sulfides, orpiment and realgar (both forms of arsenic sulfide), and pyrite (iron sulfide). None of the alkaline earths, however, are found as sulfides.

On the other hand, the alkaline earths often are found in nature as sulfates (SO_4^{2-})—unlike transition and post-transition metals, which are much less commonly found as sulfates. Familiar examples of alkaline earth sulfates include the minerals gypsum (calcium sulfate), epsomite (magnesium sulfate), and barite (barium sulfate).

THE DISCOVERY AND NAMING OF ALKALINE EARTH METALS

Lime (CaO), gypsum ($CaSO_4$), limestone ($CaCO_3$), and calcite (another form of $CaCO_3$) were all known to, and used by, ancient people. However, the chemical compositions of these substances were not known until the 19th century. By 1808, English chemist Sir Humphrey Davy had already produced by electrolysis pure samples of sodium and potassium. It was only natural that in that year he would apply the same electrolytic technique to the isolation and identification of strontium, barium, calcium, and magnesium—the major alkaline earth metals.

Davy's contemporary, Swedish chemist Jöns Jakob Berzelius (1779–1848), was working on the same problem. Berzelius wrote Davy that he had mixed lime and baryta (a mineral containing barium) with mercury, electrolyzed the mixture, and obtained *amalgams* of mercury with calcium and barium, but that he was unsuccessful at isolating the pure metals themselves. Davy repeated Berzelius's experiment mixing mercuric oxide (HgO) with lime. He was then successful in *distilling* the mercury from the amalgam of calcium and mercury that had formed, leaving behind impure metallic calcium. It was still a century later, however, before a procedure was developed to obtain pure samples of calcium.

Davy prepared barium by the same method. Minerals containing barium could be recognized by the green color their salts impart to flames. Because the mineral barite ($BaSO_4$) has a high density, the name *barium,* meaning "heavy," was chosen for that element.

In 1787, a previously unknown mineral was found in a lead mine at Strontian in Scotland; the new mineral was called "strontianite" after its place of discovery. When strontianite was dissolved in hydrochloric acid, it was found that the salt that was formed possessed chemical and physical properties very similar to the salt prepared by dissolving baryta in hydrochloric acid, and, in fact, intermediate between the properties of calcium salts and barium salts. Similar to the manner in which barium salts impart a green color to flames, it was found that strontium salts impart a red color to flames. In the same year that Davy isolated calcium and barium, he applied the same electrolytic technique to the isolation of samples of strontium metal. Just as the mineral strontianite had been named for its place of discovery, so was the name for the element, "strontium," chosen for the place of discovery.

In 1618, the water from a waterhole in Epsom, England, was found to have medicinal value. Epsom soon became the site of a spa to which visitors from all over Europe came for medical cures. Only later was the curative agent discovered to be a *hydrate* of magnesium sulfate ($MgSO_4 \times 7H_2O$), now known as *Epsom salt.* In 1808, following his isolation of calcium, barium, and strontium, Davy isolated very small samples of magnesium metal. Davy recognized that the names "magnesium" and "manganese" were so similar that it would be easy to confuse the two. To avoid confusion, he therefore recommended that the name instead be "magnium." Davy's choice of name never caught on. Ironically, if it had caught on, it probably would have spared many mistakes on tests made by subsequent generations of chemistry students!

The mineral beryl ($Be_3Al_2(SiO_3)_6$) was known to the ancient Egyptians. In 1798, French mineralogist René-Just Haüy (1743–1822) showed that beryl and emeralds have identical chemical compositions. In 1828, samples of beryllium were prepared independently by German chemist Friedrich Wöhler (1800–82) and by French chemist Antoine-Alexandre-Brutus Bussy (1794–1882) by heating beryllium chloride with metallic potassium. Wöhler obtained beryllium as a grayish-black powder. The name "beryllium" comes from the mineral beryl. Wöhler was unable to melt the beryllium, but he succeeded sufficiently in fusing the powder to make pieces of beryllium large enough to be able to observe its metallic *luster.* It was not until 1898, however, that pure beryllium was finally produced electrolytically. Bussy continued to investigate methods for preparing alkaline earth metals. In 1831, three years after his isolation of beryllium, he succeeded in preparing fairly large samples of metallic magnesium.

In the early 1900s, the Curies discovered radium, and that story is covered in chapter 9.

5

Beryllium

Beryllium is element number 4, a silvery metal with a density of 1.85 g/cm³. Besides radium, beryllium is the alkaline earth element found in the smallest quantity in Earth's crust, ranking only number 47 in relative abundance. While lithium is the lightest metal, beryllium is the second lightest metal and less chemically reactive than lithium, making it more suitable as a structural material. Because beryllium is so lightweight—but as strong as steel—it finds uses in lightweight structural materials, such as in aircraft engines, instrument parts, and structural components. Beryllium's properties more closely resemble those of aluminum than those of magnesium and the other alkaline earths. Consequently, some authors discuss beryllium together with aluminum rather than with the alkaline earths.

In this chapter, the reader will learn about the production of beryllium in stars, the forms in which it is found on Earth, its toxicity to

THE BASICS OF BERYLLIUM

Symbol: Be
Atomic number: 4
Atomic mass: 9.012182
Electronic configuration: $1s^2 2s^2$

T_{melt} = 2,349°F (1,287°C)
T_{boil} = 4,480°F (2,471°C)

Abundances:
In Earth's crust 2 ppm
In seawater negligible

Beryllium		
2	**Be**$_4$	1287°
2		2471°
	+2	
	9.012182	
	2.38×10^{-9}%	

Isotope	Z	N	Relative Abundance
$^{10}_{3}$Be	3	7	19.9%
$^{11}_{3}$Be	3	8	80.1%

humans, its chemistry, and the use of beryllium in applications such as lightweight structural materials, nuclear reactors and weapons, and alloys.

THE ASTROPHYSICS OF BERYLLIUM

The beryllium 8 isotope is produced in the collapse of massive red-giant stars via the *triple-alpha process*. Initially two helium nuclei—or alpha particles—fuse to form an unstable and very short-lived state of beryllium, as expressed by the following reaction:

$$^{4}_{2}He + {}^{4}_{2}He \rightarrow {}^{8}_{4}Be.$$

The beryllium thus produced is almost instantaneously destroyed via fission or converted to carbon by the capture of an alpha particle in the following reaction:

$$^{8}_{4}Be + {}^{4}_{2}He \rightarrow {}^{12}_{4}C + \gamma,$$

where γ is a high-energy photon or gamma ray. So helium fusion in stars cannot be the source of beryllium that is observed in stellar spectra or the interstellar medium today.

Beryllium 9—the only stable isotope of beryllium—is a product of collisions of ambient carbon, nitrogen, and oxygen molecules with cosmic ray protons. This means that beryllium 9 should be fairly evenly distributed throughout the universe when compared with iron, for example, which is strictly made in stars and ejected into space via supernova explosions of stars. Iron nuclei are heavy and would tend to remain gravitationally bound within the galaxy in which they were born. As new stars form, they gather any atomic gases in the vicinity, including ambient beryllium and iron nuclei. Comparing the abundance of beryllium to that of iron in stars can give scientists an idea of the relative ages of stellar generations: More iron in the spectrum of a star would mean more supernova events must have occurred in the galaxy before the star formed, so a star with a higher Fe/Be ratio would be younger than a star in the same galaxy that had less iron. This effect has earned beryllium the reputation as a "cosmochronometer," providing a cosmological record of time.

In the spectra of some very old stars, astronomers have observed up to 1,000 times more beryllium than expected, however. This cannot be a result of cosmic ray activity, but may be a clue to the behavior of primordial matter during the big bang. The standard theory of the big bang, which assumes a *homogeneous* explosion, provides for at most 1 percent beryllium production. If the big bang distributed matter unevenly, however, that would have allowed additional production of beryllium 9 in the lower density regions. Professor Alan Guth of the Massachusetts Institute of Technology has proposed that such a change in the distribution of primordial matter could have resulted from an exotic subatomic phase change during the earliest moments of the universe. Scientists may be able to directly test this hypothesis when the world's largest particle accelerator, the Large Hadronic Collider near Geneva, Switzerland, begins experiments designed to re-create conditions only microseconds after the big bang. According to Dr. Steven Goldfarb, a senior research fellow on the project, "We are looking for the building blocks of the universe. We want to know: What are the basic particles we are built from? And what are the rules that predict how these particles interact?"

BERYLLIUM ON EARTH

Beryllium on Earth is found in composite minerals such as granite and volcanic rock. Forms suitable for mining are beryllium silicate ($Be_3Al_2Si_6O_{18}$), bertrandite ($Be_4Si_2O_7(OH)_2$), phenacite (Be_2SiO_4), and chrysoberyl ($BeAl_2O_4$). Beryllium silicate, or beryl, is the most abundant of the three, with its most productive source found in Juab County, Utah. The gemstones emerald and aquamarine are precious forms of beryl that crystallized in cooling *magma* beneath Earth's surface. The most impressive emerald crystal (7,000 carats) that has been found to date was discovered in 1969 in Colombia, the world's largest producer of emeralds.

Beryllium is also present in solid and liquid fossil fuels. The burning of coal and oil emits beryllium particulates that may continue to circulate in the air for some time or settle into waterways. In lakes and ponds, the beryllium generally settles harmlessly to the bottom, but may enter drinking water reservoirs via rivers and streams, and has shown a marked abundance in *Precambrian* aquifers. In drinking water, levels exceeding the Environmental Protection Agency's maximum contaminant level of 4 parts per billion are extremely rare, but are most likely to occur near areas where industries dump wastewater.

Yellow vegetables such as carrots and corn tend to have higher, though nonhazardous, levels of beryllium than other vegetables. Since ingested beryllium leaves the body within hours or days, there is no health risk from this source. Inhaled beryllium, however, may take years to leave the body and can have devastating effects on the lungs. People at risk for beryllium inhalation include workers in beryllium processing plants and tobacco smokers, since beryllium tends to concentrate in tobacco. Health problems in personnel who worked with beryllium were noticed as early as the 1930s in Germany, Italy, and the Soviet Union. Beryllium pulls electrons away from the nitrogen atoms in proteins, destroying the function of body tissues. Long-term inhalation may result in *berylliosis*—scarring of the lung tissue, which can cause shortness of breath and sometimes heart problems.

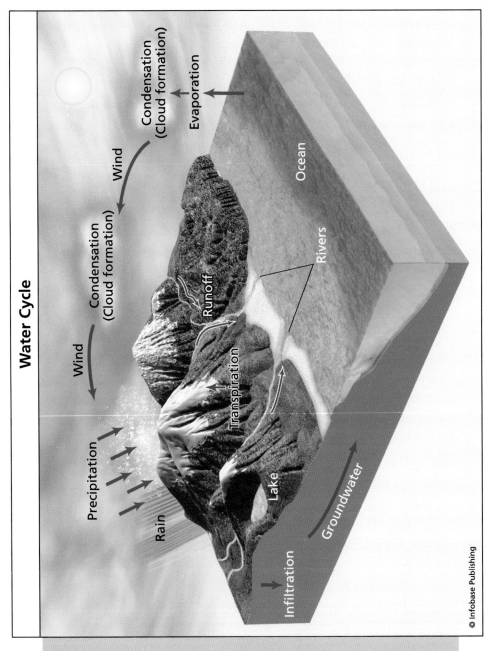

Water Cycle

The hydrologic cycle demonstrates how beryllium particulates in air may enter the groundwater.

THE CHEMISTRY OF BERYLLIUM

An element's *oxidation state* is a description of its chemical bonding. Metals tend to form positively charged ions. The charge on a simple metal ion is that element's oxidation state. Similarly, the number of covalent or ionic bonds an element has made to atoms of other elements can describe its oxidation state. A +1 oxidation state means either that the atom has only one bond to another atom or that the atom is a +1 ion. A +2 states means two bonds or a +2 ion. Like all of the alkaline earth metals, beryllium is found in compounds only in the +2 oxidation state, but it differs significantly from the other alkaline earths. In compounds containing magnesium, calcium, strontium, barium, and radium, the bonding is ionic. In compounds containing beryllium, the bonding is covalent. The prevalence of covalent bonding is one of several ways in which beryllium is similar to aluminum in its properties. (Other similarities are that both metals are lightweight and tend to resist corrosion.)

Beryllium reacts much less readily with water—even boiling water—than does its neighboring alkali metal, lithium. At low temperatures, the surface of beryllium metal forms a protective oxide coating. Beryllium dissolves in both acids and bases. Both reactions result in the evolution of hydrogen gas. In strong acids, such as hydrochloric acid, the following reaction takes place:

$$Be\ (s) + 2\ HCl\ (aq) \rightarrow BeCl_2\ (aq) + H_2\ (g).$$

In strong bases, such as sodium hydroxide, the following reaction occurs:

$$Be\ (s) + OH^-\ (aq) + H_2O \rightarrow Na^+\ (aq) + HbeO_2^-\ (aq) + H_2\ (g).$$

Beryllium oxide (BeO) is an extremely hard substance, much more so than the oxides of the other alkaline earths. Beryllium hydroxide $(Be(OH)_2)$ is an *amphoteric* compound, meaning that it is soluble in both acidic and basic solutions. In hydrochloric acid, the following reaction takes place:

$$Be(OH)_2\ (s) + 2\ HCl\ (aq) \rightarrow BeCl_2\ (aq) + 2\ H_2O\ (l).$$

(continued on page 64)

REDUCING THE CRITICAL MASS IN NUCLEAR WEAPONS

In most nuclear weapons and reactors, the fission of uranium 235 creates the heat energy needed for power production. While ^{235}U is radioactive with a half-life of about 700 million years, the natural rate of decay is slow, and alpha particles are emitted rather than neutrons that provide the chain reaction needed for nuclear power generation. Neutron-producing fission can be stimulated, however, by bombarding ^{235}U with slow or *thermal* neutrons. (Thermal neutrons have energies less than 0.4 electron volts [eV].) In the following reaction, the neutron-induced fission of ^{235}U provides energy as well as more neutrons:

$$n_{thermal} + {}^{235}U \rightarrow {}^{90}Sr + {}^{143}Xe + 3n + \text{energy.}$$

Each reaction produces about 200 million electron volts of energy. The neutrons ejected from the breakup are then available to stimulate more fission events in a chain reaction process—an ideal situation for power production. The problem is

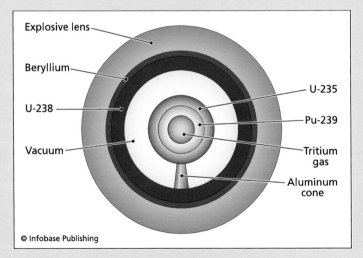

© Infobase Publishing

Beryllium is used to form a reflecting shell around the active core of a nuclear device.

that these neutrons are highly energetic. High-energy neutrons produce other reactions than the one above. They, too, can stimulate fission of ^{235}U, but the result is a high number of unusable gamma rays and very few neutrons.

The solution is to incorporate a material called a *moderator*—a material that is able to slow neutrons by collisions without absorbing them. Graphite and heavy water are often used as moderators in reactors, but for high-temperature applications like weapons, beryllium is ideal. Because beryllium does not absorb slow neutrons, the metal is also used to form a reflecting shell around the active core of the device. Neutrons within the core move in all directions. Those moving radially outward would eventually travel out of the reaction region and be absorbed or lost. The beryllium shell serves as a barrier that scatters the thermal neutrons back into the uranium-rich core, allowing them to stimulate more fission reactions. Keeping more thermal neutrons in play means more fission reactions occur, which effectively lowers the mass of uranium needed for *criticality*.

For high-temperature applications like nuclear weapons, beryllium is an ideal moderator. *(AP photo/NASA)*

(continued from page 61)

In sodium hydroxide, the following reaction occurs:

$$Be(OH)_2 \text{ (s)} + 2\,OH^- \text{ (aq)} \rightarrow BeO_2^{2-} \text{(aq)} + 2\,H_2O \text{ (l)}.$$

The beryllium ion does not form stable compounds with the carbonate ion (CO_3^{2-}), with sulfate (SO_4^{2-}), or with hydride (H^-), although it does form halides such as beryllium chloride ($BeCl_2$) and beryllium bromide ($BeBr_2$). However, unlike most metal halides, which are good conductors of electricity in the molten state, beryllium halides are not. The difference can be explained by the differences in bonding in the two groups of compounds. In most metal halides, the bonding is ionic; in the molten state, the ions are free to conduct an electrical current. In beryllium halides, however, the bonding is covalent; there are no ions to conduct a current.

The uniqueness of the beryllium ion's properties can be attributed to its very small size compared to the sizes of the ions of the other alkaline earths. Because of the small size of a beryllium atom, the valence electrons are held very tightly to the nucleus, effectively preventing the formation of a positive ion.

BERYLLIUM IS IMPORTANT IN PARTICLE ACCELERATORS

Physicists study subatomic particle interactions using particle accelerators that may be either linear or circular. The configuration chosen depends on the type of particle to be accelerated and the goal of the research. In both types, particles are generally accelerated by changing electric fields and steered by carefully calculated magnetic fields. In general, the goal of the steering process is to maintain a small particle-beam diameter and to keep the beam from hitting the walls of the pipe in which it travels before it reaches its target.

When it reaches the target, which may be an intersecting beam (a stream of gas), a liquid, or a solid, the particles that result from the collision fly off in all directions. Physicists must detect these in order to understand what happened at the moment of impact. The detector may be inside the beam pipe with electronic connections to a computer on the outside, but sometimes the detector is too large or too sensitive to conditions like temperature or *electrical conductivity* to be contained in the accelerator pipe.

Beryllium windows are often used in particle beam accelerators.
(Alexander Tsiaras/Photo Researchers, Inc.)

What is needed is a window that does not change the nature of the subatomic particles created in the collision. Beryllium is the element of choice, as it has a low atomic number, which means there are fewer protons and electrons to interfere electromagnetically. The element can be made into a metal more easily than hydrogen or lithium, and it is strong, even as a thin foil. This is important because accelerator beam pipes must be evacuated of air particles as much as possible to minimize unwanted collisions. So a huge pressure differential exists between the inside and outside of the beam pipe. Implosion would be a risk with a weaker material.

TECHNOLOGY AND CURRENT USES OF BERYLLIUM

Beryllium's light weight makes it valuable as a structural material, and it is incorporated into a number of alloys: When mixed with copper, for example, it results in a much stronger form of bronze. Beryllium copper alloy is used in computer motherboards and contact circuits. Because beryllium is extremely light and strong, its alloys with steel are used in aircraft and missiles. Also, beryllium is very stiff and withstands vibration, a property that is advantageous in acoustics: Pure beryllium *tweeters* are used in high-fidelity audio equipment. Beryllium has good *thermal conductivity,* facilitating its use as a thermal barrier to protect spacecraft from the heat buildup that occurs upon reentry into Earth's atmosphere. The pure metal is used in aircraft disc brakes, space vehicle instruments, satellite structures, missile parts, and navigational systems. Aviation applications of alloys include aircraft frames, electrical connectors, precision instruments, and engine parts. Other applications of alloys include automotive electronics, telecommunications equipment,

Beryllium-steel alloys have many uses in building construction. *(Olexa/Shutterstock)*

computers, golf clubs, and bicycle frames. Beryllium's nuclear properties make it the material of choice for windows in X-ray tubes and particle accelerators.

Several compounds of beryllium have important applications. The most commercially important beryllium compound is beryllium oxide (BeO), which is used in high-temperature applications, such as *crucibles,* microwave ovens, ceramics, and insulators. Beryllium oxide also finds use in gyroscopes and military vehicle armor. Beryllium chloride ($BeCl_2$) is used as a catalyst in the synthesis of organic chemicals. Beryllium hydride (BeH_2) is a source of hydrogen gas when mixed with water. Beryllium carbide (Be_2C) is a source of neutrons in nuclear reactors.

While beryllium is needed for these many applications, it is costly. In nature, beryllium does not exist in metallic form; it must be processed by humans, who are susceptible to the risks of inhalation. At the time of this writing, no new processing facilities are planned in the United States, largely because of the expense involved in conforming to environmental standards. The Brush Wellman Company in Utah is still the world's largest producer of beryllium, however, providing more than 90 percent of world needs, and their mines continue to be highly productive.

6

Magnesium

Magnesium, element number 12, is a hard, lustrous, silvery metal with a density of 1.74 g/cm³; it is one of the most widely distributed elements on Earth, ranking number 7 in relative abundance. Like the other alkaline earths, the pure element is never found in nature, and all magnesium compounds are in the form of the Mg^{2+} ion. There are two major sources of magnesium: seawater and the mineral magnesite ($MgCO_3$). In the production of sodium chloride (table salt, NaCl), when seawater is allowed to evaporate, magnesium salts can be recovered along with the sodium salts. Magnesite, found in the United States mainly in Colorado, is also a common source of magnesium. Magnesium is essential to both plants and animals as a component of *chlorophyll* in green plants, bones in animals, and numerous *enzymes* in a variety of living organisms. In addition, magnesium helps build proteins and facilitates DNA replication.

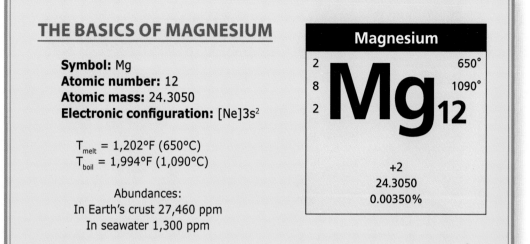

THE BASICS OF MAGNESIUM

Symbol: Mg
Atomic number: 12
Atomic mass: 24.3050
Electronic configuration: [Ne]3s^2

T_{melt} = 1,202°F (650°C)
T_{boil} = 1,994°F (1,090°C)

Abundances:
In Earth's crust 27,460 ppm
In seawater 1,300 ppm

Magnesium	
2	650°
8	1090°
2	**Mg** 12

+2
24.3050
0.00350%

Isotope	Z	N	Relative Abundance
$^{24}_{12}$Mg	12	12	78.99%
$^{25}_{12}$Mg	12	13	10.00%
$^{26}_{12}$Mg	12	14	11.01%

In this chapter, the reader will learn about the nucleosynthesis of magnesium in stars; the recovery of magnesium on Earth; the chemistry of magnesium, including its role in photosynthesis; and important applications of magnesium, including its use in superconductors.

THE ASTROPHYSICS OF MAGNESIUM

The mechanisms by which magnesium is synthesized in stars depend on stellar properties such as mass, temperature, and density. In stars massive enough that the core reaches a temperature where carbon burning (the fusion of two carbon atoms) can occur, two different magnesium isotopes, ^{23}Mg and ^{24}Mg, can result from the following reactions:

$$^{12}_{6}C + {}^{12}_{6}C \rightarrow {}^{23}_{12}Mg + {}^{1}_{0}n$$

$$^{12}_{6}C + {}^{12}_{6}C \rightarrow {}^{24}_{12}Mg + \gamma,$$

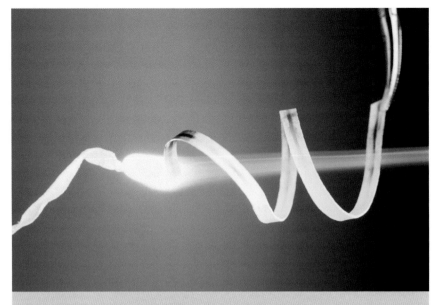

Magnesium burns in air to produce a very brilliant white light.
(Charles D. Winters/Photo Researchers, Inc.)

where *n* represents a neutron and γ a high-energy photon. Carbon burning can also produce oxygen nuclei, which, if the star uses up all the lighter elements in its core, can subsequently fuse to form magnesium 24:

$$^{16}_{8}O + ^{16}_{8}O \rightarrow ^{24}_{12}Mg + 2^{4}_{2}\alpha,$$

where α represents an alpha particle or helium nucleus. This proceeds only in very massive stars where the central density can rise to at least 10^{10} kg/m^3.

The end of life for these stars is cataclysmic, but not as completely destructive as a supernova explosion. A star four to eight times as massive as the Sun typically supports not only a carbon-burning core, but also a helium-burning layer as well as a hydrogen-burning shell. This is an unstable situation that leads to an explosive ejection of nearly all of the light gases that comprise the outer atmosphere of the star and a sort of implosion of the core. What remains is a strange object whose state of matter is impossible on any planet or even within the Sun. In this state, electrons can and must exist in a higher density than is possible in an atom. Such exotic astrophysical objects are called *white dwarfs*. They are not really stars, as fusion has stopped. They shine only because of infrared radiation—heat—and continue to fade over the centuries.

By contrast, a supernova explosion completely obliterates a star, expelling all its matter into space to be collected by other stars that continue the evolution of nucleosynthesis. The explosion releases a blast of alpha particles that can interact in myriad ways to produce elements built up by nuclear multiples of four. One such nucleus is magnesium 24, which is built when alpha particles collide with neon nuclei during the explosion:

$$_{10}^{20}\text{Ne} + \alpha \rightarrow {}_{12}^{24}\text{Mg} + \gamma$$

$$_{10}^{21}\text{Ne} + {}_{2}^{4}\alpha \rightarrow {}_{12}^{24}\text{Mg} + {}_{0}^{1}\text{n}.$$

One isotope of magnesium does not rely on such exotic processes as those described here, but the process slowly builds up over thousands of years in stellar atmospheres via alpha capture as follows:

$$_{10}^{22}\text{Ne} + {}_{2}^{4}\alpha \rightarrow {}_{12}^{25}\text{Mg} + {}_{0}^{1}\text{n},$$

mainly in old high-mass stars, called AGB (asymptotic giant branch) stars. These stars shed their elements into the atmosphere during thermal pulsations, which are poorly understood but are effective at providing the interstellar medium with the heavy elements now found on Earth.

MAGNESIUM ON EARTH

Magnesium makes up about 2.4 percent of *Earth's crust*. Magnesium is the lightest metal that can be used for structural materials. (There are only a few metals lighter than magnesium, and two of them—lithium and sodium—would react with any environmental moisture.) By itself, magnesium would not be strong enough to support heavy weights, but it is strengthened when alloyed with other metals.

Magnesium is mined in significant quantities from the minerals dolomite ($CaMg(CO_3)_2^4$) and magnesite ($MgCO_3$) and in smaller quantities from brucite ($Mg(OH)_2$) and carnallite ($KMgCl_3 \times 6(H_2O)$).

Magnesium salts are abundant in natural mineral brines—which result mainly from evaporated saltwater—but are also an oil field by-product. Magnesium can be extracted from saltwater areas such as the Great Salt Lake, the Dead Sea, and the oceans. The Dead Sea has the highest concentration of magnesium at 3.4 percent by weight. On average, ocean water contains only about a tenth of a percent, but, being a vast supply and available to all coastal areas, provides a renewable source of magnesium salts.

Certain ocean precipitates incorporate the Mg^{2+} ion in their makeup. In calcite ($CaCO_3$), it has been shown that the magnesium content enhances calcite solubility, which, in turn, slows crystal growth. This raises concerns about increasing the magnesium content of seawater (as from oil production), which can have adverse implications for *calcareous* marine organisms such as *plankton*.

Recent research on *Earth's mantle* indicates that magnesium may provide a clue to some of the most baffling behavior of the material comprising the region 1,700 miles (2,700 km) beneath Earth's surface. Scientists have long pondered how a material that must be solid can actually flow like a *viscous* substance, and, even more surprisingly, display unexpected changes of speed and direction. Laboratory experiments that simulate the extremely high pressures of the Earth's interior now show the possibility of a new phase of a magnesium silicate ($MgSiO_3$) that bonds in two directions rather than three. In an April 2006 issue of *Nature,* Japanese research scientist Kei Hirose remarked that "(it is) like a stack of two-dimensional sheets." This new phase of matter could indeed flow and help drive convection, and may explain minute changes in Earth's rotation.

THE CHEMISTRY OF MAGNESIUM

Magnesium metal is prepared by electrolysis of *molten* magnesium chloride ($MgCl_2$). In this process, $MgCl_2$ is heated until it melts, and an electrical current is passed through it. Molten magnesium metal is collected at the cathode, while chlorine gas is generated at the anode, as shown in the following reaction:

$$MgCl_2 \ (l) \rightarrow Mg \ (l) + Cl_2 \ (g).$$

As happens in the electrolysis of other salts, the chlorine gas that is produced is an important source of chlorine for water purification systems.

When magnesium burns in air, it produces a brilliant white light. In the chemical reaction that takes place, the product is magnesium oxide, as shown by the following equation:

$$2 \text{ Mg (s)} + \text{O}_2 \text{ (g)} \rightarrow 2 \text{ MgO (s)}.$$

Because this reaction is highly exothermic, magnesium can be used in *incendiary* bombs.

Magnesium metal is used in signal flares and fireworks; as the metal burns, it increases their brilliance. In the recent past of flash photography—when flashbulbs could be used only once—batteries supplied an electric current that passed through a small piece of magnesium, producing the flash.

Magnesium oxide ("magnesia") has an extremely high melting point—4,788°F (2,642°C)—and is used in the making of magnesium carbon bricks for the walls of furnaces where the ability to withstand high temperatures is essential. The magnesium ion (Mg^{2+}) also forms stable compounds with common anions. Examples include magnesium halides (MgF_2, $MgCl_2$, $MgBr_2$, and MgI_2), magnesium nitrate ($Mg(NO_3)_2$), magnesium carbonate ($MgCO_3$), magnesium sulfate ($MgSO_4$), magnesium phosphate ($Mg_3(PO_4)_3$)), magnesium oxalate (MgC_2O_4), and magnesium hydroxide ($Mg(OH)_2$). The fluoride, carbonate, phosphate, and hydroxide are insoluble in water; the other compounds are soluble. In qualitative analysis, the insolubility of $Mg(OH)_2$ can be used to separate Mg^{2+} from Ca^{2+}, Sr^{2+}, and Ba^{2+}, since the hydroxides of the latter three are fairly soluble.

In nature, the compound magnesium sulfate is recovered from the mineral epsomite, which is used to produce epsom salt, $MgSO_4 \times 7H_2O$. Epsom salt is sold in pharmacies for use as a laxative. It can also be used as a soaking aid for sore feet or sprained ankles.

Magnesium is an essential element for both plants and animals. Green plants manufacture their own food in the presence of sunlight because plants contain chlorophyll, which itself contains the magnesium ion. Magnesium ions are also required in the replication of DNA and RNA molecules in the cells of all living organisms. All chemical reactions that take place in living organisms are catalyzed by enzymes. Magnesium ions activate the enzyme molecules.

Magnesium is one of the metals (along with zinc, for example) that are used in *cathodic protection* of steel. Steel structures that are underground or that are used in the construction of boats tend to corrode in the presence of moisture and dissolved oxygen. Because magnesium is a more active metal than iron is, placing a block of magnesium in contact with steel causes the magnesium to corrode instead. This protects the

MAGNESIUM-DIBORIDE SUPERCONDUCTORS

Material science is an ancient study, but one that offers tantalizing possibilities to this day. The simple *binary compound* magnesium diboride (MgB_2), for example, which is easily and inexpensively fabricated by exposing magnesium wire to boron gas in an otherwise evacuated chamber, has currently come into the limelight. In 2000 C.E., it was discovered that MgB_2 is superconducting at temperatures below 39 K. This means that electrons traveling through the MgB_2 lattice structure encounter virtually no resistance to their motion. It seems that, in this material, electrons—which would normally repel each other—are able to pair up by exchanging energy provided by lattice vibrations.

While 39 K is extremely cold, it is a warmer temperature than other materials need to become superconducting, and it can be achieved by standard refrigeration techniques rather than expensive *cryogenic* cooling. Wires made of superconducting MgB_2 are most useful in magnet windings that carry high current, since they will not suffer the heat (resistance) losses that are proportional to the current squared (as in regular copper wire, for example). This material is especially promising in the medical application of magnetic resonance imaging, which requires high magnetic fields for visualization of hydrogen atoms in human physiology. Ultrahigh magnetic fields will allow visualization of other atoms such as carbon and sodium, allow-

steel from corrosion and increases the lifetimes of underground tanks and the hulls of ships. Essentially, what takes place is an electrochemical process referred to as *galvanic action* (similar to what takes place in a commercial battery). In a galvanic cell, oxidation occurs at the anode (negative electrode) and reduction at the cathode (positive electrode). In the absence of a protective metal like magnesium or zinc, iron would serve as the anode and would oxidize. However, by placing a metal in contact with iron that is more active than iron, the more active metal becomes the anode and is oxidized (hence the term *sacrificial anode* to

Members of the MgB_2 research team gathered at the National Synchrotron Light Source at Brookhaven National Laboratory in New York. *(Brookhaven National Laboratory)*

ing more accurate diagnosis and prognosis of many neurological diseases. Promising research indicates that higher magnetic fields can be applied to superconducting wires without exceeding the *critical current density* if carbon atoms are incorporated in the wire material.

refer to the metal's role), while the iron (or steel) becomes a cathode. Since neutral metals or alloys cannot be reduced, the iron itself remains inert, while dissolved oxygen or other environmental substances are, in fact, being reduced. The magnesium or zinc is sacrificed, but replacing a block of one of those metals is easier and cheaper than replacing the structure they are protecting.

THE HEART OF CHLOROPHYLL

Chlorophyll supplies the green pigmentation of all Earth's vegetation, from simple algae to the most massive of trees. The heart of the molecule is the Mg^{2+} magnesium ion. Chlorophyll is a *chelate,* which simply means that a metal ion is an integral and central component binding a large organic molecule. In the case of chlorophyll, the organic molecule is a *porphyrin,* comprised of carbon, hydrogen, nitrogen, and oxygen.

Chlorophyll supplies the green pigmentation of all Earth's vegetation, from simple algae to the most massive of trees. *(Michael Shake/Shutterstock)*

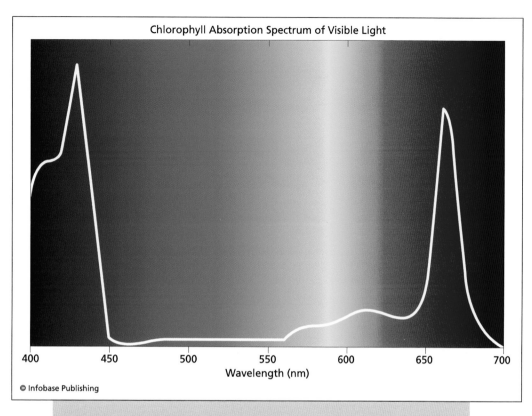

Chlorophyll Absorption Spectrum of Visible Light

Wavelength (nm)

© Infobase Publishing

The chlorophyll absorption spectrum shows the lack of absorption in the green wavelengths.

Amazingly, this structure is identical to that of *hemoglobin,* with the only difference being that hemoglobin has an iron ion at its core.

Chlorophyll's major importance lies in its role as the mediator of photosynthesis, the process by which plants utilize the Sun's energy to produce food for growth. Chlorophyll absorbs sunlight to initiate energy transfer. The absorption range is in the red and blue-violet edges of the visible spectrum; plants appear green because of reflected light. Most plants can transform light into electrical and then chemical energy with an impressive 95 percent efficiency—a fact that has prompted research into using chlorophyll-like substances in solar electricity cells. Sunlight promotes the electrons in chlorophyll from lower- to higher-energy quantum states of the molecule. Only recently have these excitations been understood to occur under

Chlorophyll—a molecule diagram showing magnesium ion at the core

resonant conditions that allow coherent coupling among electronic states. The coherent nature of the excitations allow for nearly instantaneous transfer of energy to neighboring electrons, which significantly speeds the subsequent construction of energy-storing molecules. The energy is then used to assist the transfer of electrons from water to carbon dioxide in order to form carbohydrates for food.

$$CO_2 + H_2O \rightarrow CH_2O + O_2$$

German chemist Richard Willstätter (1872–1942) was awarded the 1915 Nobel Prize in chemistry for his exhaustive experimentations on chlorophyll and photosynthesis. He showed that magnesium was crucial to the photosynthetic process and that, while composed of two slightly different molecules called chlorophyll-a and chlorophyll-b, the substance is the same in all green plants. He also designed a method to produce chlorophyll in the laboratory in quantities sufficient for experimental research.

TECHNOLOGY AND CURRENT USES OF MAGNESIUM
Magnesium is combined with aluminum, zinc, manganese, zirconium, and other metals in a number of alloys, where its light weight and high

strength-to-weight ratio are beneficial. These alloys are used primarily in the aircraft and transportation industries, where light weight saves fuel. They are also used in racing cars, where the lighter weight increases speed. Magnesium alloys are also used in portable power tools, cameras, office machines, sporting goods, and luggage.

Magnesium is used to protect structures made of steel. Moisture, atmospheric oxygen, and other chemicals in the environment that would degrade steel react with magnesium instead, protecting the underlying steel structure. The metal is also used in the production of less reactive metals from their ores. Magnesium metal reacts strongly with oxygen and burns to produce heat and white light, which makes it useful for signal flares, fireworks, and incendiary bombs.

A number of magnesium compounds have useful applications. These compounds include the acetate, *alkyls,* borate, boride, bromide, carbonate, chloride, hydroxide, nitrate, oxide, peroxide, phosphates, silicide, stannide, stearate, and sulfate. Magnesium acetate ($Mg(C_2H_3O_2)_2$) is used in the production of rayon fiber and as a deodorant, disinfectant, and antiseptic. Magnesium alkyls (where an *alkyl* is a fragment of a hydrocarbon molecule such as $-CH_3$ or $-CH_2CH_3$) are used to produce plastics and detergents. Magnesium boride (MgB_2) is a superconducting material with potential uses in magnets and electronics.

Magnesium borate ($3MgO \times B_2O_3$) is used as an antiseptic and as a *fungicide.* Magnesium bromide ($MgBr_2$) is recovered from seawater and is the principal compound used to manufacture liquid bromine (Br_2). In medicine, $MgBr_2$ is used as a sedative. Magnesium carbonate ($MgCO_3$) is used as an antacid and as an additive to ink, glass, and cosmetics. $MgCO_3$ is also added to table salt to prevent "caking" that would occur due to absorption of atmospheric moisture. Magnesium chloride ($MgCl_2$) is the major magnesium salt recovered from seawater. A brine consisting of magnesium chloride is used to melt ice on roads and to suppress dust at construction sites, mines, and unpaved parking lots. $MgCl_2$ is also used in water treatment and fire extinguishers.

Magnesium hydroxide ($Mg(OH)_2$) is used in the pulp and paper industries. In medicine, it is used as an antacid and a laxative. Magnesium nitrate ($Mg(NO_3)_2$) may be used as a fertilizer to provide a source of nitrogen. Magnesium oxide (MgO, also known as *magnesia*) is used in water treatment, the paper industry, household cleaners, ceramics for aircraft windshields, fertilizers (to promote photosynthesis), semiconductors, and food additives. Magnesium peroxide (MgO_2) is a bleaching agent. There are several magnesium phosphates ($Mg(H_2PO_4)_2$, $MgHPO_4$, and $Mg_3(PO_4)_2$). These can be used in the production of plastics, and as antacids, food additives, nutritional supplements (as a source of Mg^{2+}), and dentifrice polishing agents.

Magnesium silicide (Mg_2Si) and magnesium stannide (Mg_2Sn) are used in the semiconductor industry. Magnesium stearate ($Mg(C_{18}H_{35}O_2)_2$) is used in medicines and cosmetics. Magnesium sulfate ($MgSO_4$) is used in animal feed and fertilizers. In the form $MgSO_4 \times 7H_2O$, it is referred to as *Epsom salt* and is used as a laxative and a soaking agent.

Worldwide magnesium production is more than 800,000 tons (725,748 metric tons) annually. As demand for lightweight automobile components grows to increase fuel efficiency, demand for magnesium should continue to increase.

7

Calcium

Calcium, element number 20, is a silvery-gray metal with a density of 1.55 g/cm³. The main source of calcium is seawater. Sedimentary materials that contain calcium include calcite, limestone, marble, and chalk (all of which are forms of calcium carbonate, $CaCO_3$), gypsum ($CaSO_4 \times 2H_2O$), and dolomite ($CaMg(CO_3)_2$).

Calcium-containing compounds are ubiquitous in the natural environment. As a consequence, they should be familiar to students in beginning chemistry and physical geology courses. $CaCO_3$, for example, is an important component of corals and shells. Sedimentary rocks often contain the fossil remains of these marine organisms, and the presence of $CaCO_3$ in these rocks is easily detected. Geologists carry dropper bottles of dilute hydrochloric acid (HCl) into the field. If they suspect that a rock contains $CaCO_3$, they will add a few drops of HCl to

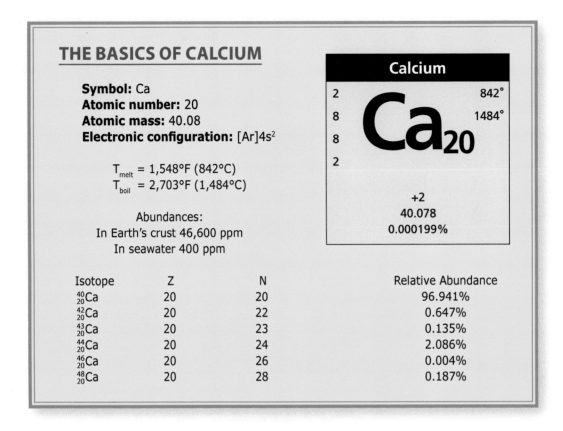

THE BASICS OF CALCIUM

Symbol: Ca
Atomic number: 20
Atomic mass: 40.08
Electronic configuration: $[Ar]4s^2$

T_{melt} = 1,548°F (842°C)
T_{boil} = 2,703°F (1,484°C)

Abundances:
In Earth's crust 46,600 ppm
In seawater 400 ppm

		Calcium		
2				842°
8		**Ca**$_{20}$		1484°
8				
2				
		+2		
		40.078		
		0.000199%		

Isotope	Z	N	Relative Abundance
$^{40}_{20}Ca$	20	20	96.941%
$^{42}_{20}Ca$	20	22	0.647%
$^{43}_{20}Ca$	20	23	0.135%
$^{44}_{20}Ca$	20	24	2.086%
$^{46}_{20}Ca$	20	26	0.004%
$^{48}_{20}Ca$	20	28	0.187%

it. If $CaCO_3$ is present, the rock will fizz due to the evolution of carbon dioxide gas, as shown by the following equation:

$$CaCO_3 \text{ (s)} + 2 \text{ HCl (aq)} \rightarrow CaCl_2 \text{ (aq)} + CO_2 \text{ (g)} + H_2O \text{ (l)}.$$

Teachers commonly use this reaction to demonstrate acid-base chemistry.

Calcium is well known to most people because it is an essential component of teeth and bones. A deficiency of calcium can lead to *osteoporosis,* which results in low bone density and an increased risk of bone fracture. Soy products, green vegetables, and dairy products are important sources of calcium and therefore are part of a healthy diet.

In this chapter, the reader will learn about the nucleosynthesis of calcium in stars, how the minerals from calcium are recovered, the chemistry of calcium, the importance of calcium in the structure of teeth and bones, and applications of calcium and its compounds.

Green leafy vegetables are important sources of calcium and, therefore, part of a healthy diet. *(Olga Lyubkina/Shutterstock)*

THE ASTROPHYSICS OF CALCIUM

The number of nuclei comprising calcium's core is in a multiple of four, which makes it an alpha-process element. This means that calcium is generally synthesized in stars by the collisional capture of a helium nucleus (α) onto an argon 36 atom, as represented in the following:

$$^{36}_{18}\mathrm{Ar} + {}^{4}_{2}\alpha \rightarrow {}^{40}_{20}\mathrm{Ca}.$$

A curious circumstance, however, is that calcium appears to be about half as abundant as other alpha-process elements in galaxies. The causes are not clear, but observations indicate that calcium abundance in stars is directly correlated with the mass of the star and the velocity variation within the star at the time of formation of the calcium nuclei. Further studies of supernovae, with their complex velocity distributions, should inform theories of calcium nucleosynthesis.

Type-1 supernovae (SneI) are of particular interest in the study of calcium production: This type of supernova occurs in binary star systems where one of the members is a white dwarf (see chapter 5) and the other is a normal *main sequence* star. In such a system, the stars revolve around their common center of mass. The dense white dwarf gravitationally attracts gases in the outer atmosphere of the companion star. Those accelerating gases become extremely hot as they accelerate toward the surface of the white dwarf and finally ignite in a thermonuclear blast so powerful that both stars are destroyed, flinging their matter into space. One such supernova that was observed in 2001 showed a strange feature in its emission spectrum—an extremely high-velocity ionized calcium jet moving at twice the speed and with a different *polarization* from another, more typical jet in the same explosion. The explosion in the case of this SneI is hypothesized not to have been spherically symmetric, as it exhibits differing polarization angles of the light emitted. Scientists speculate that it actually blew off clumps of matter rather than gasifying everything. It will be important to observe numerous supernovae spectra to determine more certainly what processes occur in the various types.

Focusing back in on objects in our own galaxy, particularly the Sun, astrophysicists continue to study spectra of solar prominences in order to measure the calcium/hydrogen ratio—one indicator of gas pressure in the Sun's atmosphere. Doubly ionized calcium (Ca II or Ca^{2+}) happens to emit visible light with a wavelength of 854 nanometers, which was first photographed by American astronomer George Hale (1868–1938), using his newly invented spectroheliograph in 1889. He found that calcium was not distributed evenly around the Sun but instead occurred in "clouds" in certain areas. Since then, it has been observed that these areas move and change. Observations also show that the relative density of Ca II in meteor spectra is quite high, at about 10^{14} particles per cubic centimeter. Ionized calcium gas in the Sun's atmosphere most likely results from meteors and comets traversing circumsolar space. As they pass, particles stream from their tails and are *sublimated* by the Sun's energy. Models that study the temperature dependence of calcium ionization imply that the atom is most easily ionized at temperatures below 6,000 K, which occur near the surface of the Sun.

CALCIUM ON EARTH

Comprising nearly 3.5 percent of Earth's crust, calcium is found in igneous, sedimentary, and metamorphic rock. The most commonly known form, calcium carbonate (also known as calcite or $CaCO_3$), is found as the sedimentary mineral limestone, although there are impurities that are almost never mentioned in the chemical formula. Marble, the fabric of famous sculptures, is a product of Earth's extreme heat and pressure on limestone, which forces it to metamorphose into its beautiful crystalline structure. Calcites can also have a dramatic effect on volcanic activity. Magma flowing over limestone heats the rock and breaks the bonds in the calcite, resulting in pockets of carbon dioxide. The CO_2 bubbles can quickly increase the volume of the magma, leading to an eruption.

Calcium carbonate is currently more actively studied in aqueous solution. Earth's oceans are saturated with Ca^{2+} ions, especially near the surface. The calcium is delivered to the oceans mostly by rivers and streams. (As rocks weather, calcium ions are released and attracted to water molecules.) Because there is such a high concentration of calcium ions in the oceans, it easily precipitates out as calcium carbonate or, less commonly, as calcium sulfate. The precipitate contributes to the formation of coral reefs in warm lagoons and shallow tropical seas. Inland salt seas and lakes, such as the Dead Sea in Israel, the Great Salt Lake in Utah, and Mono Lake in eastern California, often have whitish limestone deposits around their shores. (All of these bodies of water are saltier than the oceans.) Softer deposits tend to be called chalk. In locations where spring water bubbles up through the floor of a salty lake, mounds of limestone, called *tufas,*

Marble, like that used for the U.S. Supreme Court building, is a product of Earth's extreme heat and pressure on limestone. *(Gary Blakeley/iStockphoto)*

may form. Mono Lake is especially known for its picturesque tufa formations.

Calcite is also crucial to aquatic organisms like oysters, claims, snails, corals, and sea urchins that use it to build their shells. As these shellfish die, their shells add to the coral reefs and to the layers of limestone already building up on the ocean floor. Eons later, the sea floor rises and the limestone layers become geological features of valleys, canyons, and mountains. Because $CaCO_3$ is an important component of corals and shells, these sedimentary materials often contain the fossil remains of once-living organism.

THE CHEMISTRY OF CALCIUM

At ordinary temperatures, calcium metal is stable in the air in the absence of moisture. When heated, the metal reacts with atmospheric oxygen to form white calcium oxide, as shown in the following equation:

$$2 \text{ Ca (s)} + \text{O}_2 \text{ (g)} + \text{heat} \rightarrow 2 \text{ CaO (s)}.$$

The metal burns with a very bright light, and the fumes are a skin and eye irritant.

In the presence of moisture, the metal readily reacts to form calcium hydroxide and hydrogen gas, as shown in the following equation:

$$Ca\ (s) + 2\ H_2O\ (l) \rightarrow Ca(OH)_2\ (s) + H_2\ (g).$$

Bonding in calcium compounds is always ionic. The calcium ion, Ca^{2+}, forms a large number of compounds, including the following:

- calcium nitrate $(Ca(NO_3)_2)$
- calcium halides $(CaF_2,\ CaCl_2,\ CaBr_2,\ and\ CaI_2)$
- calcium carbonate $(CaCO_3)$
- various calcium phosphates $(Ca_3(PO_4)_2,\ Ca(H_2PO_4)_2,\ CaHPO_4,$ and other more complex compounds)
- calcium oxide (CaO)
- calcium hydroxide $(Ca(OH_2))$
- calcium nitride (Ca_3N_2)
- calcium phosphide (Ca_3P_2)
- calcium sulfite $(CaSO_3)$, and
- calcium sulfate $(CaSO_4)$.

The carbonate and phosphates are insoluble in water. The nitrate and halides are soluble. The remaining compounds tend to have intermediate solubilities.

Limestone can form in two ways: with or without the aid of living organisms. In the inorganic mechanisms, the chemical reactions that form calcium carbonate begin with the solution of atmospheric carbon dioxide (CO_2) in water. Dissolved CO_2 reacts with water to form carbonic acid (H_2CO_3), the familiar acid in carbonated soda, which in turn reacts with water to form bicarbonate ions (HCO_3^-). These reactions are represented by the following equations:

(1) $$CO_2\ (g) \leftrightarrow CO_2\ (aq);$$

(2) $$CO_2\ (aq) + H_2O\ (l) \leftrightarrow H_2CO_3\ (aq);$$

(3) $$H_2CO_3\ (aq) + H_2O\ (l) \leftrightarrow HCO_3^-\ (aq) + H_3O^+\ (aq),$$

where the double-arrow symbol ($\leftrightarrow$) indicates that the reactions can go in both directions. Bicarbonate ions can then react to form carbonate ions, as shown by the following fourth equation:

$$(4) \qquad 2\,HCO_3^- \,(aq) \leftrightarrow CO_3^{2-} \,(aq) + CO_2 \,(aq) + H_2O \,(l).$$

The ratio of HCO_3^- to CO_3^{2-} depends on the acidity of the solution—the more basic the solution, the more CO_3^{2-} that forms. Ca^{2+} ions are in the water naturally from dissolved minerals. Once CO_3^{2-} forms, Ca^{2+} and CO_3^{2-} combine and precipitate as calcium carbonate, as shown by the following fifth equation:

$$(5) \qquad Ca^{2+} \,(aq) + CO_3^{2-} \,(aq) \leftrightarrow CaCO_3 \,(s).$$

The precipitation of $CaCO_3$ is facilitated by several mechanisms. For example, if the water evaporates, the concentrations of the ions increases until the water is saturated; at any higher concentrations, precipitation occurs. If the temperature increases, the water holds less dissolved CO_2; the result is a shift in the equilibrium in equation (4) that favors an increase in both CO_2 and CO_3^{2-}. A third mechanism is wave action. Along the shores of lakes and oceans, wave action releases CO_2 to the atmosphere, again shifting the equilibrium in equation (4) to increase the concentration of CO_3^{2-} ions.

This series of reactions illustrates the mechanism by which a number of limestone formations develop in nature. For example, in a cave, groundwater saturated with Ca^{2+} and HCO_3^- ions seeps through the cave ceiling to form hard deposits of $CaCO_3$ called *stalactites*. Some of the seepage will drip to the floor, building up columns of $CaCO_3$ called *stalagmites*. Sometimes the two formations will grow together, forming a solid column from floor to ceiling.

Havasu Canyon in Arizona—home to the Havasupai Indian Tribe—has beautiful waterfalls in the middle of the Arizona desert. The same reactions that form stalactites and stalagmites are responsible for the travertine formations (a form of calcium carbonate) coating the cliffs under the falls that contribute to the beauty of the falls. Havasu Creek has a high lime content that replenishes the travertine—the lime gives the water a greenish tinge. Because Havasu Creek is the lifeblood of the

(continued on page 90)

IN MORTAR AND BONES

People have burned limestone to make lime ($CaCO_3$) for the construction industry for millennia. At least as far back as 7000 B.C.E., the mixing of lime in mortar had become a chemistry experiment—one that continues to this day.

Mortar (sometimes called cement) is used to bond surfaces like bricks together, but also for plastering walls. Historically, it has been composed variously of lime, sand, clay, volcanic rock and ash, brick dust, and potsherds. Early lime mortars that set simply by reaction between the lime and carbon dioxide in the air offered little protection from deleterious effects of water to the structure. Aggregate mortars that incorporate *pozzolans* and silicates, which react to bond with calcium, do not need CO_2, and some can even set underwater. These are called hydraulic mortars, and offer durability in weather, but are less suitable for situations where *plasticity* is needed, as in restoration projects, for example.

(continues)

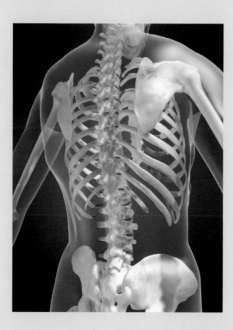

As with mortar and concrete, bone strength is maintained by calcium, though it is still unclear what the optimum is in dietary supplementation. *(Sebastian Kaulitzki/ Shutterstock)*

(continued)

Cement production is a huge business; about 2 billion tons (1.8 metric tons) are produced annually, and this is predicted to rise. Unfortunately, this industry is currently responsible for as much as 10 percent of the world's emissions of carbon dioxide, mostly from the burning of lime. The problem has prompted research into the replacement of calcium with some other easily available material. One possibility is magnesium, as it bonds similarly with silicates.

Porosity and *permeability* are important considerations in mixing the best mortar for a particular project. There is a certain ratio of strength to weight that makes porosity so relevant. It is much the same concern with bone structure. As with mortar and concrete, bone strength is maintained by calcium (though it is still unclear what the optimum is in dietary supplementation). Researchers that synthesize bone for hip and knee replacements are looking at new ceramics that are porous without being fragile. Getting just the right combination of calcium and phosphate compounds seems to be key. Ideally, they will formulate a material that bonds well with healthy bone cells and is easily manufactured and machined.

(continued from page 88)

tribe, it gives the Indians their name. "Havasupai" means "People of the blue-green waters."

CALCIUM IMAGING OF THE BRAIN

Doubly ionized calcium (Ca^{2+}) delivers information about the stars, about processes in Earth's oceans, and even about how the human brain sends messages through the central nervous system. Because the ions carry excess charge, they move in response to changing voltages. When a neuron fires in the brain, it instantaneously changes the electric potential between the interior and exterior of the cell, allowing a current of calcium ions to mediate the message.

The relationships among calcium concentration, neuron signals, *synaptic* strength, and other cells of the central nervous system are extremely complex but quite important for brain metabolism. Periodic Ca^{2+} current modulation has been linked to cell growth and secretion. Interrupted or altered Ca^{2+} signaling may be connected to diseases such as Alzheimer's and long-term depression.

Scientists have recently developed a way to track calcium ions in a living brain by bonding them with *fluorescent* molecules that emit visible light. The light can then be detected using advanced spectroscopic techniques. This process, called calcium imaging, allows the observation of actual communication between neurons, and has already led to an increased understanding of brain circuitry. The 2008 Nobel Prize for chemistry was awarded to Japanese-American chemist Osamu Shimomura (1928–) and American chemists Martin Chalfie (1947–) and Roger Y. Tsien (1952–) for their work in the development of green fluorescent protein, which allows similar tracking of proteins and enzymes, and may lead to insights regarding the behavior of cancer cells.

TECHNOLOGY AND CURRENT USES OF CALCIUM

As the third most abundant metal in Earth's crust, calcium is widespread in a large number of mineral deposits, relatively inexpensive to recover, and useful in a number of applications. In industry, calcium is used in the refining of metals like lead, aluminum, zirconium, and uranium. Calcium alloyed with iron in steel production reduces surface defects. When alloyed with lead in the manufacture of maintenance-free automobile batteries, it increases battery life. The metal is also used to produce vitamin B_5, calcium pantothenate.

Calcium is found in rocks as the carbonate, fluoride, oxide, phosphate, and sulfate. In purified forms, each of these minerals has practical applications. Limestone, calcite, and marble occur in the form of calcium carbonate ($CaCO_3$) and are used as building materials. In addition, marble is carved to make statues. Portland cement—derived mainly from calcium carbonate and calcium silicates—is a staple of the building industry. Calcium carbonate is taken to relieve heartburn; antacids like

Tums® are pure $CaCO_3$. Calcium fluoride (CaF_2) is the principal source of fluorine in the production of hydrofluoric acid (HF). Hydrofluoric acid is often the source of fluorine in the processes of manufacturing fluorine-containing chemicals. The mineral fluorite (CaF_2) occurs in nature in several colors and is used as a gemstone.

Calcium oxide (CaO also known as lime) is used in metallurgy in the reduction of iron alloys to iron metal. Lime is also used in water treatment and to manufacture cement. As is the case with magnesium, calcium phosphates exist in several forms—$Ca(H_2PO_4)_2$, $CaHPO_4$, and $Ca_3(PO_4)_2$. Dental products often contain calcium phosphate materials, since teeth themselves are made of calcium phosphate. $Ca_3(PO_4)_2$ is added to spices to prevent caking. It is also used in fertilizers. $Ca(H_2PO_4)_2$ is added to baking powder and plant food. $CaHPO_4$ is added to animal food. Calcium sulfate ($CaSO_4$) occurs in nature as the mineral gypsum and is used to make wallboard. Another form, plaster of Paris, is beloved by artists and designers who use it to cast small art objects or work wonders with interior walls. Alabaster is made of gypsum and can be carved to make sculptures.

Other calcium compounds include calcium chloride ($CaCl_2$), which can be spread on sidewalks and roads as a deicer in winter; calcium hypochlorite ($Ca(ClO)_2$), which is a form of powdered bleach; calcium nitrate ($Ca(NO_3)_2$), which is a fertilizer; calcium carbide (CaC_2), which is the primary ingredient used to produce acetylene in miners' lamps; and calcium sulfide (CaS), which is used sometimes as a hair remover.

More than 280 million tons of calcium (as CaO) are produced worldwide on an annual basis. With calcium's many versatile uses, calcium products can be expected to be in demand for many years to come.

8

Strontium
and Barium

Strontium, element number 38, has a density of 2.63 g/cm³ and is the 16th most abundant element on Earth. Barium, element number 56, has a density of 3.51 g/cm³ and ranks 14th in abundance. Both elements are silvery-colored metals. Because alkaline earths react so readily with any water in the environment to form ions and compounds, neither element would ever be found as the *native metal*. In all of their ores, they occur as +2 ions. The principal sources of the two elements are the minerals celestite ($SrSO_4$), strontianite ($SrCO_3$), and barite ($BaSO_4$).

Under natural conditions, neither element has any biological importance. In some cases, however, strontium can be of concern in regards to human health. Strontium is perhaps best known for its radioactive isotope Sr-90, which possesses a half-life of 26 years. Strontium 90 is

THE BASICS OF STRONTIUM

Symbol: Sr
Atomic number: 38
Atomic mass: 87.62
Electronic configuration: $[Kr]5s^2$

T_{melt} = 1,431°F (777°C)
T_{boil} = 2,520°F (1,382°C)

Abundances:
In Earth's crust 360 ppm
In seawater 8.1 ppm

	Strontium	
2		777°
8	**Sr**$_{38}$	1382°
18		
8		
2		
	+2	
	87.62	
	7.7×10^{-8}%	

Isotope	Z	N	Relative Abundance
$^{84}_{38}$Sr	38	46	0.56%
$^{86}_{38}$Sr	38	48	9.86%
$^{87}_{38}$Sr	38	49	7.00%
$^{88}_{38}$Sr	38	50	82.58%

THE BASICS OF BARIUM

Symbol: Ba
Atomic number: 56
Atomic mass: 137.33
Electronic configuration: $[Xe]6s^2$

T_{melt} = 1,341°F (727°C)
T_{boil} = 3,447°F (1,897°C)

Abundances:
In Earth's crust 340 ppm
In seawater 0.03 ppm

	Barium	
2		727°
8	**Ba**$_{56}$	1897°
18		
18		
8		
2	+2	
	137.327	
	1.46×10^{-8}%	

Isotope	Z	N	Relative Abundance
$^{130}_{56}$Ba	56	74	0.106%
$^{132}_{56}$Ba	56	76	0.101%
$^{134}_{56}$Ba	56	78	2.417%
$^{135}_{56}$Ba	56	79	6.592%
$^{136}_{56}$Ba	56	80	7.854%
$^{137}_{56}$Ba	56	81	11.232%
$^{138}_{56}$Ba	56	82	71.698%

one of the many products of uranium fission. It is part of the waste products from nuclear reactors and part of the radioactive fallout from atomic explosions.

Barium is probably best known in the form of barium sulfate ($BaSO_4$). Barium sulfate is a white solid that is insoluble in the human *gastrointestinal (GI) tract.* It is administered in powdered form as a contrast agent for use in X-ray pictures of the upper and lower GI tracts. Any abnormal growths—cancerous *tumors,* for example—will clearly show as dark spots in the X-rays against the white background of the $BaSO_4$.

In this chapter, the reader will learn about the nucleosynthesis of strontium and barium, the occurrence of both elements on Earth, the chemistry of both elements, and their practical applications.

THE ASTROPHYSICS OF STRONTIUM AND BARIUM

Interstellar strontium and barium, both being heavier than iron, are synthesized in supernovae via the rapid capture by iron nuclei of a succession of neutrons—the r-process. The barium and strontium in old stars in the outer portions of galaxies (sometimes called "thick-disk" stars) probably synthesized via the r-process, as the proportion of heavy elements in that more primitive interstellar medium owed its existence almost solely to Type II supernovae.

Some fraction of heavy alkaline-earth elements also built up slowly over thousands of years in the atmospheres of large-mass stars via neutron capture and electron release, with the requirement that iron 56 nuclei—remnants of prior supernova explosions—be available as seeds. Because this synthesis proceeds relatively slowly due to a low density of neutrons, it is called the s-process.

The s-process can also occur in the helium-burning inner shells of low-mass asymptotic giant branch (AGB) stars that evolve through the red-giant stage. AGB stars have cores containing carbon and oxygen nuclei surrounded by a helium shell and an outer hydrogen shell that is characterized by enormous thermal pulsations. These pulsations occur when the underlying helium layer begins to fuse into carbon, quickly increasing the star's *luminosity* and generating a high-velocity stellar wind that ejects massive amounts of stellar material, including strontium and barium, into space. Meteorites contain approximately the same percentage of barium in their composition as low-mass AGB stars.

However, scientists have observed some stars that have an unusually high abundance of barium, strontium, and heavier elements—a phenomenon that cannot be explained by analysis of individual star evolution. These strange objects are called "barium stars," not because they are made of barium, but because of the unexpected overabundance of that element in their spectra. The most likely way a star could acquire such a wealth of barium, strontium, and other heavy s-process elements is by capturing the gases from another nearby star. This is not an unexpected situation, since the majority of stars in the universe revolve in binary star systems. Most known barium stars are red-giant stars that have gravitationally attracted the matter blown off a companion AGB star that shed its gases during thermal pulsations.

Astrophysical strontium production, too, has shown itself to be of extreme interest, especially as regards a particular stellar spectrum, that of the exceedingly luminous star Eta Carinae, which exploded magnificently. It was presumed to be a supernova that would have destroyed the star completely. But it did not. The nebula that formed after the explosion could be the focus of study for a long time. There is a feature called the strontium filament, rich in strontium, that is a curiosity. Its very existence has generated a new way of thinking about supernovae. Supernovae are known to produce large quantities of neutrinos as well as their antiparticles, called antineutrinos. Antineutrinos can change a proton into a neutron, and may play a major role in the building of rare isotopes during stellar explosions.

STRONTIUM AND BARIUM ON EARTH

Both strontium and barium are found on Earth in layers of sedimentary rock as a result of eons of the elements' precipitation out of seawater to the ocean floors. Barite ($BaSO_4$), a soft but remarkably heavy inert and insoluble mineral, is mined throughout the world, mainly for use in oil and gas fields. In order to avoid "gushers" (the explosive release of pressure in oil and gas wells), workers prefill the hole with a barite mud whose weight damps the upward force of the gusher. The similar mineral compound celestite ($SrSO_4$) could also serve this purpose, but it is more expensive. There are currently nine barite mines in the United States, but no mines for the extraction of celestite. A problem with both of these materials is that they tend to leave a residue. Hard deposits form within the well and in the machinery used to extract the oil. This is rough on the equipment and can cause it to malfunction.

The west coast of North America may contain the only known streaks of sanbornite, a rare barium silicate ($BaSi_2O_5$) that was listed as a new mineral in the early 1930s, but only investigated in detail in the

Barite is the most common of barium ores. *(Bureau of Mines, Mineral Specimens)*

decades following World War II. Since the discovery, more than seven new barium silicates have been named from this interesting metamorphic deposit.

Although strontium is less abundant in Earth's crust than barium, it plays an important role in geological, archaeological, and climatological research. The isotopic ratio $^{87}Sr/^{86}Sr$ in seawater, for example, is remarkably uniform regardless of sample location. When something happens to change that ratio—such as tectonic plate activity, continental weathering, or hydrothermal activity—it takes less than 1,000 years (a very brief period on geological timescales) for ocean circulation to equalize the distribution once again. Oceanographers and geologists have, therefore, been able to construct a *calibration curve* plotting the strontium ratio versus time as far back as 200 million years. An archaeologist attempting to date a stone tool made from sedimentary rock or a climatologist needing the age of a particular piece of coral can have the sample measured for its $^{87}Sr/^{86}Sr$ content. Comparing that with the calibration curve can give a very accurate idea of the age of the piece. The strontium isotope content can also determine the origin of formation. In 2007, scientists from the Max Planck Institute for Evolutionary Anthropology in Leipzig used strontium isotope dating of tooth enamel to conclude that some Neanderthals traveled widely, which had been debated by other anthropologists.

THE CHEMISTRY OF STRONTIUM AND BARIUM

Like the other alkaline earth elements, strontium and barium are never found in pure form in Earth's crust. They react readily with water, all of the halogens, and oxygen. Their chemistry is dominated by the +2 ions—Sr^{2+} and Ba^{2+}.

The chemical compounds these ions form are very similar to each other. In addition, they are very similar to the compounds formed by calcium and, to a lesser extent, to the compounds formed by magnesium. Most of the compounds of the alkaline earth ions are white solids and are commonly encountered in high school and college chemistry laboratories (and are much more commonly found than compounds of beryllium or radium—the former because of its toxicity and the latter because of its scarcity and radioactivity). Several of these compounds are summarized in the following table:

	NO_3^-	Cl^-	OH^-	O^{2-}	CO_3^{2-}	SO_4^{2-}	PO_4^{3-}
Mg^{2+}	$Mg(NO_3)_2$	$MgCl_2$	$Mg(OH)_2$	MgO	$MgCO_3$	$MgSO_4$	$Mg_3(PO_4)_2$
Ca^{2+}	$Ca(NO_3)_2$	$CaCl_2$	$Ca(OH)_2$	CaO	$CaCO_3$	$CaSO_4$	$Ca_3(PO_4)_2$
Sr^{2+}	$Sr(NO_3)_2$	$SrCl_2$	$Sr(OH)_2$	SrO	$SrCO_3$	$SrSO_4$	$Sr_3(PO_4)_2$
Ba^{2+}	$Ba(NO_3)_2$	$BaCl_2$	$Ba(OH)_2$	BaO	$BaCO_3$	$BaSO_4$	$Ba_3(PO_4)_2$

As dry solids, all of the compounds in a column are very similar in appearance. In addition, they tend to have similar properties. For example, the nitrates and chlorides are all very soluble in water. However, there are some differences that may be noted. For example, the solubility in water of the hydroxides and oxides increases upon descending a column. $Mg(OH)_2$ and MgO are weak bases and insoluble in dilute aqueous solutions. Descending the columns, the strength as bases increases, so that $Ba(OH)_2$ and BaO are strong bases. The carbonates and phosphates of all four cations are insoluble in water, with the insolubility increasing upon descending a column. Finally, $MgSO_4$ is soluble in water, $CaSO_4$ only slightly soluble, and $SrSO_4$ and $BaSO_4$ both completely insoluble.

Two anions that precipitate remarkably differently with the various alkaline earth ions are oxalate ($C_2O_4^{2-}$) and chromate (CrO_4^{2-}). The solids of the four oxalates are white, but only magnesium oxalate (MgC_2O_4) is soluble. The solids of the four chromates are yellow (due to the yellow color of the chromate ion itself). $MgCrO_4$ and $CaCrO_4$ are very soluble in water. $BaCrO_4$ is very insoluble. The solubility of $SrCrO_4$ is intermediate. In a solution containing both Ba^{2+} and Sr^{2+}, adding small amounts of K_2CrO_4 will precipitate $BaCrO_4$ almost exclusively.

The differences in solubility are used in the qualitative analysis of these ions. Suppose a solution initially contains one or more of these ions: Mg^{2+}, Ca^{2+}, Sr^{2+}, and Ba^{2+}. In addition, suppose the solution could contain one or more of the alkali metal ions, especially Na^+ and/or K^+. (In practice, all other metal ions—the transition and post-transition metals—would already have been removed before testing for alkaline earth or alkali metal elements.) Adding Na_2CO_3 will precipitate the

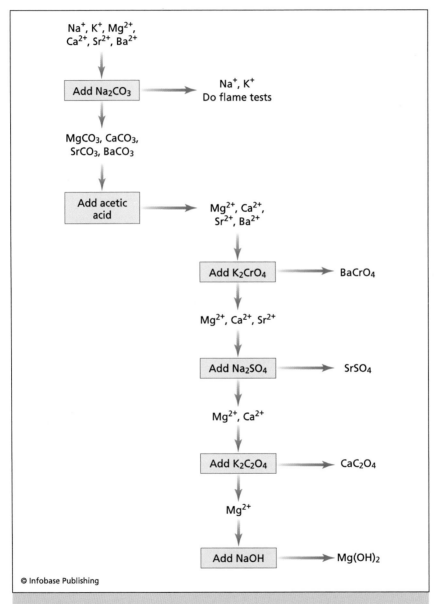

© Infobase Publishing

Flowchart of alkaline earth and alkali metal separations *(Source: Brian Nordstrom)*

A MATERIAL HARDER THAN DIAMOND

Diamond has long been known as the hardest material on Earth. Now it will have to be reclassified as the hardest naturally occurring material. In early 2007, scientists at the University of Wisconsin found that tin supplemented with barium titanate ($BaTiO_3$) proves itself harder than diamond, but only in a certain very small temperature range. They introduced tiny (0.1 mm diameter) barium titanate particles into molten tin and watched it as it cooled. They found that at around 140°F (60°C), the samples were up to 10 percent stronger than diamond. As the material cools, the barium titanate crystals become embedded in the tin's metal lattice. Free barium titanate crystals expand as they cool, but in this case they are caged by the tin. At the critical temperature, the crystals are held back from expanding and cannot undergo the phase transformation that would normally occur. In this situation, the potential to expand remains and acts to oppose external lateral forces.

The application of such an external force is the method used to test for hardness (also called stiffness). The harder the material, the more resistant it is to bending. While this new material exhibits its extreme stiffness at a rather inconvenient temperature (room temperature would be more useful), researchers are optimistic that in many situations the phenomenon could be put to good use. They are also working toward finding ways to achieve the same effect at lower temperatures.

alkaline earths that may be present, leaving behind the alkali metal ions in solution. The supernatant solution (which would contain any alkali metals present) can be decanted, separating the alkaline earths from the alkali metals. (If desired, the presence of the alkali metals can be detected using flame tests.) The solid carbonates can be washed and redissolved in dilute acetic acid.

Adding K_2CrO_4 to the resulting solution will precipitate $BaCrO_4$, leaving Mg^{2+}, Ca^{2+}, and Sr^{2+} behind in solution. The supernatant can be decanted. Adding Na_2SO_4 to the supernatant solution will precipitate $SrSO_4$, leaving Mg^{2+} and Ca^{2+} behind in solution. The supernatant can be decanted. Adding $K_2C_2O_4$ to the supernatant will precipitate CaC_2O_4, leaving Mg^{2+} in solution and confirming the presence of Ca^{2+} in the original solution. The supernatant can again be decanted. Adding NaOH will precipitate any Mg^{2+} ions that are present, confirming the presence of Mg^{2+} in the original solution. (Alternatively, sometimes $Mg_3(PO_4)_2$ is precipitated.) These steps are shown in the flowchart on page 100.

These reactions provide an example of the type of work that takes place in assays of minerals. In the mining industry, raw ores are extracted from the ground. Chemists who work for the mining companies must be able to decompose ores into their component elements. In doing so, they assay or analyze the minerals to determine two things: What elements are in them (a technique known as *qualitative analysis*), and what percentage of an ore is a given element (a technique known as *quantitative analysis*)? The flowchart of alkaline earth and alkali metal separations presented on page 100 is an example of qualitative analysis.

TECHNOLOGY AND CURRENT USES OF STRONTIUM AND BARIUM

There are relatively few uses of elemental strontium and barium. The primary use for barium metal is to remove oxygen gas from the inside of electronic vacuum tubes. Barium and strontium can replace lead in glass making, with a lower danger of toxicity to the worker. Otherwise, applications of these two elements make use of their compounds.

In fireworks, barium nitrate ($Ba(NO_3)_2$) imparts the color green. Strontium chloride ($SrCl_2$) imparts unique shades of red and purple for which there are no substitutes. Barium nitrate is also used in infrared flares, and various strontium and barium compounds are used in small-arms tracers.

An important barium compound is barium sulfate ($BaSO_4$). Barium atoms are highly X-ray and gamma-ray absorbent, a quality that is in demand in the medical industry, where it is especially useful as a diagnostic tool for diseases of the *gastrointestinal tract*—the stomach and the upper and lower intestines. Tissues in the stomach and intestines will not normally show up on X-ray film. X-rays taken after a patient swallows barium sulfate powder mixed with water will show the track of the barium through the system. If a cancerous tumor or other abnormal growth is present, it will image as a dark object against the white $BaSO_4$ background.

Barium and strontium sulfate are used as weighting agents in oil wells to preserve the structural integrity of the walls of a well. Barium sulfate, together with zinc sulfide (ZnS) and zinc oxide (ZnO), is a pigment for white paints.

Uses for the carbonates of the two elements are completely dissimilar. Barium carbonate ($BaCO_3$) is used in rat poison. Barium ions are actually toxic to humans and other animals. The barium sulfate used in medical diagnosis, however, is so insoluble that none is absorbed by the body. With rat poison, however, the carbonate reacts with the rat's stomach acid, releasing barium ions into the rat's body, thereby overstimulating heart muscle contractions and killing the rat. Strontium carbonate ($SrCO_3$) has a much more benign use: It provides the red color in color televisions.

Barium and strontium have been incorporated in new materials research. Barium titanate ($BaTiO_3$) is used in ceramic *capacitors* and, when incorporated in tin, makes a material harder than diamond. Crystals of barium titanate are also used in optical materials. Strontium titanate ($SrTiO_3$) has such a high refractive index that it sparkles like diamond and is used in less expensive jewelry.

Strontium isotopes can be used to determine the ages of rocks. Radioactive rubidium 87 decays into strontium 87, with a half-life of 4.9 billion years. Geologists can use the ratio of Sr-87 to naturally occurring

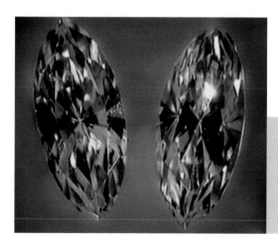

Strontium titanate has a high refractive index that enables it to sparkle like an actual diamond. It is frequently used in less expensive jewelry. *(theimage.com)*

Sr-86 to estimate the age of a rock. Because 4.9 billion years is older than the age of the Earth, this ratio provides geologists with an estimate of the age of the Earth.

Another strontium isotope, Sr-90, is highly radioactive, with a half-life of 29 years. Strontium 90 has been used as a power source in weather stations and buoys.

Most uses of strontium and barium are fairly stable. The biggest increase in production is expected to be of strontium carbonate because it provides the color red to color televisions, the numbers of which are increasing exponentially worldwide.

9

Radium

Radium is element number 88, in which all of its isotopes are radioactive; hence, what little radium is found on Earth is mostly as a trace element in uranium ores. The most common isotope has a mass number of 226 with a half-life of 1,604 years. The second longest-lived isotope is radium 228, with a half-life of 5.77 years. The other isotopes have much shorter half-lives ranging from microseconds to days. Radium is constantly being formed as part of the radioactive decay series of uranium and thorium. Because it decays so quickly, however, only minute quantities of radium ever exist at any one time.

An important part of radium's legacy is the fact that it was discovered through very painstaking work by the famous husband-and-wife scientific team of Pierre and Marie Curie. In this chapter, the reader will learn of the Curies' work. The reader will also learn about the chemistry and uses of radium and how radium is formed in Earth's crust.

THE BASICS OF RADIUM

Symbol: Ra
Atomic number: 88
Atomic mass: All isotopes are
  radioactive.
Electronic configuration: [Rn]7s^2

T_{melt} = 1,292°F (700°C)
T_{boil} = 2,084°F (1,140°C)

Abundances:
In Earth's crust trace amount
In seawater negligible

	Radium	
2		700°
8		
18	**Ra$_{88}$**	
32		
18		
8	+2	
2	[226]	

Isotope	Z	N	Half-life
$^{223}_{88}$Ra	88	135	11.43 days
$^{224}_{88}$Ra	88	136	3.64 days
$^{226}_{88}$Ra	88	138	1,604 years
$^{228}_{88}$Ra	88	140	5.77 years

RADIUM ON EARTH

Radium is synthesized along with many other heavy elements by rapid neutron capture in supernovae explosions, but since all radium isotopes are radioactive and decay to other elements, virtually none of the radium found on Earth is of cosmic origin. The radium isotope with the longest half-life (1,604 years) is ^{226}Ra, which is a product of the spontaneous fission of ^{238}U. Radium 223, with a half-life of 11.4 days, is a daughter of ^{235}U, while the fission of ^{232}Th produces both ^{228}Ra and ^{224}Ra, with half-lives of 5.8 years and 3.6 days, respectively. Basically, anywhere uranium and thorium are found in Earth's crust, radium will also exist.

In general, radium in rock does not pose a health risk, but there are situations where radium is found in unusual abundance, and care must be taken with human exposure. Oil fields, coal mines, and phosphate mines generate hazardously high concentrations of radium as uranium and thorium deposits are disturbed underground. In Poland, for example, radium-containing water that was discharged from coal mines has been responsible for the contamination of rivers. As discussed in chapter 8, barium mineral mud is injected into oil-field drill holes because its weight can inhibit gushers. But this mud commonly contains radium crystallized within the barite. When the sludge moves through machinery, it leaves radioactive deposits that stay there. Care must be taken with waste procedures to ensure that the public is not exposed to the radioactive materials. Some also remains in the soil, and surface spreading can occur. This problem is difficult to avoid and nearly impossible to remove.

The issue in coal mines is less complicated; coal mines are so far underground that groundwater is full of minerals that contain radium. The mining process often channels these underwater flows to the surface, where excess radium can accumulate. Research into sulfate used in treatments has found, however, that certain natural clays preferentially absorb radium. Strategic placement of the clay can allow for collection and removal of radioactive isotopes.

Groundwater discharge of the short-lived radium isotopes ^{223}Ra and ^{224}Ra from coastal wetlands has actually proven useful to scientists who model flow rates through sediments. Measurements of isotopic ratios can provide information on vertical transport of the groundwater in shallow bays and other marshy wetlands.

The half-life of radium 226 makes its abundance in the upper layers of ocean sediments, which settled within the past 10,000 years (Holocene epoch), convenient to measure. A comparison of ^{226}Ra abundance to ^{228}Ra in various ocean locations allows for determination of ocean current directional flow: Because ^{228}Ra is produced more strongly in shallow areas, and its lifetime is so much shorter, observation of radium 228 far from shore can indicate offshore currents that are otherwise difficult to measure.

THE DISCOVERY AND NAMING OF RADIUM

All isotopes of radium are radioactive and occur naturally in very small quantities. In their studies of radioactive elements, husband and wife Pierre (1859–1906) and Marie (1867–1934) Curie discovered that pitchblende—the mineral that was their source of uranium—was more radioactive than its uranium and thorium content alone could explain. Therefore, they suspected that a new, undiscovered element was probably present in pitchblende, that it constituted about 1 percent of the pitchblende, and that it was much more intensely radioactive than either uranium or thorium. Madame Curie decided to look for this new element as the basis of her doctoral thesis.

Although Curie was a well-known and distinguished physicist, the working conditions he could offer his wife were meager. In 1898, in a cramped shed at the University of Paris—damp and freezing in winter, stifling hot in summer—the couple began a four-year collaboration to sift through tons of pitchblende, trying to isolate the new element. What the Curies found was not just one new element, but two new elements. Madame Curie named the first one they isolated "polonium" after her home country of Poland. Because the second new element they found was a "giver of light," or equivalently, a "giver of rays," the Curies called it "radium." Contrary to the Curies' prediction that these elements would constitute 1 percent of the pitchblende from which they were extracted, they discovered instead that they constituted but a millionth of the pitchblende.

Because radium was present in such a tiny amount, the Curies were at first unable to isolate pure samples, but had to settle for the chloride salt, $RaCl_2$. Much to the Curies' delight, they found that radium chloride glowed in the dark, just like phosphorus. The Curies took such joy in their discovery that they would entertain their scientific friends—among them Paul Langevin (1872–1946) and Jean-Baptiste Perrin (1870–1942)—by inviting them to the Curies' apartment just to sit in the dark and watch their sample of radium chloride glow.

In 1902, Madame Curie finally succeeded in isolating one-tenth of a gram of pure radium, a sample sufficient also to determine radium's atomic weight of 225 grams per mole. Wholesale production of radium began two years later, resulting in the isolation of the first full gram of radium. That same year, the Curies shared the Nobel Prize in physics with Henri Becquerel—the Curies for their study of radioactivity, and Becquerel for the discovery of radioactivity.

Tragedy struck the Curie family on April 19, 1906. On the streets of Paris during a rainstorm, Pierre Curie accidentally stepped in front of a moving horse-drawn wagon. Slipping on the wet pavement, he fell before the horses. He escaped being trampled by the horses, but

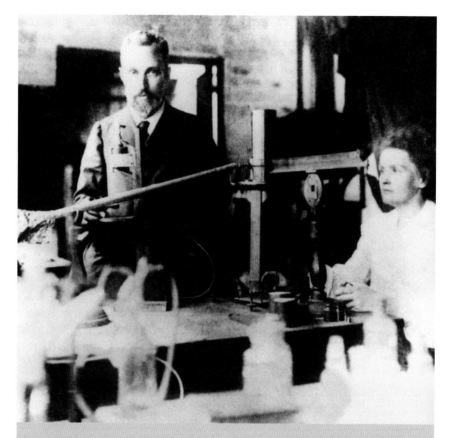

Husband and wife Pierre Curie and Marie Sklodowska Curie are shown in their laboratory in France in this undated photo. *(AP Photo)*

the left back wheel of the heavy wagon struck him, crushing his head and killing him immediately. Marie never remarried, but continued an active career in science. After her husband's death, Madame Curie also received in 1911 the Nobel Prize in chemistry for the discoveries of polonium and radium.

In the early years of working with radioactive substances, no one understood the dangers of exposure to radiation or took the precautions we take today to minimize exposure. On July 4, 1934, Marie Curie succumbed to a lifetime of radiation poisoning.

Today, Pierre and Marie Curie are remembered for their early study of radioactivity. Pierre is remembered in physics for his studies in electricity and magnetism. He discovered the *piezoelectric effect,* in which small electric currents are generated when the volume of a crystal is changed. He also discovered that some materials may be magnetic only below a certain temperature (called the *Curie temperature*). Above that temperature, the magnetism disappears. Marie is best remembered as the discoverer of radium. After Pierre's death, Marie succeeded him as the head of his laboratory at the Sorbonne in Paris. Marie spent the last two decades of her life developing the use of X-rays in medicine and finding medical applications for radioactive materials.

THE CHEMISTRY OF RADIUM

Lying just below barium in the periodic table, radium's chemistry is essentially identical to barium's chemistry. In fact, there is a famous story in which confusion between the two elements played an important role. In the 1930s, Italian physicist Enrico Fermi (1901–54) and his coworkers were investigating the action of neutrons on samples of uranium. (The neutron had only been discovered in 1930. Its use in physics was still relatively new.) Their expectation was that the absorption of neutrons by uranium would lead to the production of transuranium elements (elements lying beyond uranium in the periodic table), as shown by the following equation:

$$\underset{92}{\overset{N}{U}} + \underset{0}{\overset{1}{n}} \rightarrow \underset{92}{\overset{N+1}{U}} \rightarrow \underset{-1}{\overset{0}{e}} + \underset{93}{\overset{N+1}{X}},$$

where N = mass number of a uranium isotope, $\underset{0}{\overset{1}{n}}$ is the symbol for a neutron, $\underset{-1}{\overset{0}{e}}$ is the symbol for a beta particle (i.e., an electron), and $\underset{93}{\overset{N+1}{X}}$ indicates that element 93 has been formed.

When Fermi's group analyzed the products of the neutron bombardment, it appeared to them that radium had been produced, especially since they had no reason to even suspect that barium could be a product. Since radium is the daughter element formed by two successive alpha decays of a uranium atom, they decided their quest for a transuranium element was unsuccessful. Subsequently, Otto Hahn (1879–1968), Fritz Strassmann (1902–80), and Lise Meitner (1878–1968), all from Germany, reinterpreted the results to show that it was not radium atoms that had been formed, but barium atoms instead from the nuclear fission of uranium. Thus, Fermi and his group just missed discovering fission.

(continued on page 112)

RADIUM HOT SPRINGS

In North America, there are two places named Radium Springs (in New Mexico and Georgia) and one called Radium Hot Springs (in British Columbia). All three are named for the local springs that well up from deep within the Earth. All groundwater has some radium content, but thermal springs in general seem to have more, for various reasons. The Canadian site has been said to have the highest radioactivity rate of any thermal springs, but this cannot be known, since there are thousands of sites, and not many have actually been monitored.

Geothermal springs are often situated in volcanic regions. Vertical transport of brines from deep underground is one source of radium in such areas. It has also been found that the radium content is higher in acidic springs, such as those in Yellowstone National Park in the United States. Acidic water more easily erodes rock. If the local rock has a high uranium 238 content, erosion releases the radium 226 daughter product in gaseous form into the aquifer.

(continues)

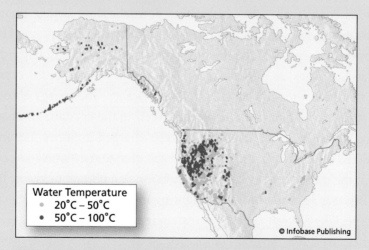

Water Temperature
- 20°C – 50°C
- 50°C – 100°C

© Infobase Publishing

Plot of the warmest thermal springs in United States. The orange dots indicate water temperatures of 70°–120°F (21.1°–48.9°C), whereas the red dots run from 120°F up to the boiling point.

(continued)

Those springs that have been the focus of scientific studies are widely dispersed. Results from Iran, Israel, Tunisia, Morocco, Taiwan, Arkansas, and Nevada indicate radium 226 levels in thermal springs to be higher than the average in public drinking water reservoirs but below the maximum contaminant level designated by the U.S. Environmental Protection Agency.

(continued from page 110)

RADIOLUMINESCENCE AND THE PAINT THAT KILLS

In the early 1900s, the hazardous nature of radioactivity was not yet understood—its novelty lay in its ability to make objects glow. The French physicist Henri Becquerel had discovered that alpha and beta particles could make some crystalline powders glow. Unlike photoluminescence, this phenomenon is not induced by the absorption of light, but by the absorption of the energy from the radioactive fission products (alphas and betas) as they are slowed down and stopped in the crystals. It is, therefore, properly called *radioluminescence.* The energy absorbed allows electrons to move around more freely in the crystals. These wandering electrons are attracted to positively charged impurity sites in the minerals. Upon arrival, they cause an excitation and eventual de-excitation at the site, which results in the emission of a photon of visible light.

This was seen to be ideal for lighting things without using electricity, especially aircraft and ships' instruments, emergency signage, and watches. By 1905, several companies combined zinc sulfide powder and radium 226 ore to make radioluminescent paint with names like "Undark" and "Marvelite." By 1920, more than 4 million clocks and watches sported dials painted with radium. Alpha particles given off by radium 226 are fairly harmless to humans when encountered in the out-

Radium paint was used for lighting things without using electricity, especially aircraft and ships' instruments, emergency signage, and watches. *(SSPL/The Image Works)*

doors or from any external source because they cannot penetrate skin. If ingested or inhaled, however, radium is extremely dangerous because its similar electronic arrangement allows it to behave like calcium in the body and accumulate in bone.

The laborers who painted the watch faces had particularly delicate work to do. The tiny numbers required a perfect tip on the brush, and it was a common practice to lick the tip to make it paint just right. This went on all day long every day of work, so workers consumed radium-infused paint at a steady rate—for years in many cases. Many of the workers, who were mostly women, eventually developed bone cancer, particularly of the jaw.

RADIOPHARMACEUTICALS—A GOOD USE OF RADIOACTIVITY

Radioactive elements with short half-lives are particularly useful for imaging and treatment of tumors. One way for radiotherapy to be delivered to the location in the body where it will do the most good is by injection of a small source or "seed" to a specific site, a process called

Workers at Radium Dial in Ottawa, Illinois, are shown in this undated photo. The town today still grapples with the tragedy of the workers who suffered as a result of hand-painting clock-face dials using radium-laced paint in the 1920s and the subsequent cleanup process after the company's buildings were demolished years later. *(AP Photo/Ottawa Daily Times)*

brachytherapy. The radiation emitted by a radioisotope is like a bullet that, if properly aimed, can damage target receptor cells, and is especially potent against those that divide rapidly like cancer cells. Some isotopes of strontium, barium, and radium are especially convenient for cancer treatments, mainly because they are easily found in nature or produced in particle accelerators.

Because they are chemically similar to calcium, radioisotopes of strontium, barium, and radium are "bone-seeking" and are key in targeting lesions of the bone in a way that inflicts minimal damage to neighboring healthy tissues. Because of their high energy and short half-life, alpha particles are ideal for this treatment, so radiologists like to choose radioisotopes with a decay chain that involves alpha particles and short half-lives. Radium 223 has been the most promising, as it has only an 11-day half-life, short half-lives of all daughter isotopes, and produces four alpha particles per decay (see the accompanying decay table for radium isotopes). It is now marketed in the chemical form $^{223}RaCl_2$ under the brand name Alpharadin®. Trials have shown good results with fortunately little uptake in bone marrow.

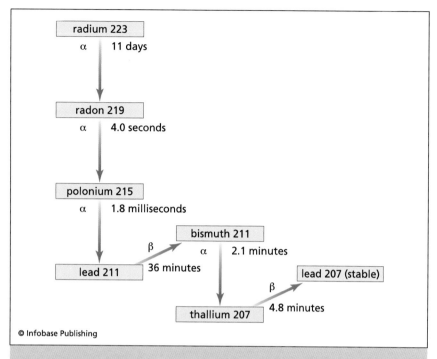

The radium 223 decay chain, showing particle emissions, daughter nuclei, and half-lives of daughter products

Other uses of alkaline earth radioisotopes include pain relief treatments. It has been found that the beta emissions of strontium 89, for example, have the effect of lessening pain in patients with bone disease, as well as in those with bone pain related to prostate, liver, and thyroid cancers.

TECHNOLOGY AND CURRENT USES OF RADIUM

Radium 223 is promising in the treatment of bone cancer, as it has only a half-life of 11 days, short half-lives of all daughter isotopes, and produces multiple alpha particles per decay. It is now marketed in the chemical form $^{223}RaCl_2$.

In some nuclear reactors, radium provides alpha particles that can interact with beryllium to produce neutrons via the following reaction.

$$^{9}_{4}Be + ^{4}_{2}\alpha \rightarrow ^{12}_{6}C + ^{1}_{0}n$$

The neutrons are then used to help maintain the crucial chain reaction process.

The use of radium for glow-in-the-dark paint cannot be relegated to history. While radium-laced paint was banned in the United States and Europe in the 1950s, the product and practice has not been prohibited in all countries.

Radium is useful to scientists who study flow rates in sediments. Groundwater discharge of the short-lived radium isotopes ^{223}Ra and ^{224}Ra from coastal wetlands allows the measurement of isotopic ratios that provide information about vertical transport of the groundwater in shallow bays and other marshy wetlands. For ocean scientists, a comparison of ^{226}Ra abundance to ^{228}Ra in various locations allows for determination of ocean current directional flow, which may prove especially useful in studies of the ocean's response to climate change.

10

Conclusions and Future Directions

The periodic table of the elements is a marvelous tool that scientists have only begun to investigate. The key to its utility is its organization, the patterns it weaves. It can guide the eye and the mind to understand how far science has come and where human knowledge can lead.

UNDERSTANDING PATTERNS AND PROPERTIES IN THE ALKALI AND ALKALINE EARTH METALS

The location of their valence electrons in s subshells imposes restrictions on the alkali metals and alkaline earths that are not imposed on elements elsewhere in the periodic table. First of all, with the exceptions of the lightest members of each family, the metals are too reactive in air and in water to use the neutral metals themselves in any applications.

Each element is limited to only a single ion (+1 for the alkali metals, +2 for the alkaline earths), with the result that *complex ion* chemistry is almost absent in these families.

The nature of the electronic configuration in the alkaline earth metals does, however, allow for intriguing research into correlations between the two electrons in the outer shell of these atoms. Simultaneous excitation of the two outer electrons is possible with a single photon, which leads to the provocative idea of a shared principal quantum number. Excitation of one of the electrons to consecutively higher orbits while monitoring de-excitation of the other electron gives information about the border between quantum and classical (planetary) behavior.

This shared electronic configuration has a more pragmatic importance in medicine. The similarity in electron shells allows strontium, barium, and radium to be absorbed in the same way as calcium in human and animal physiology—a phenomenon that leads to both a cause and a treatment for cancer.

SPECULATIONS ON FUTURE DEVELOPMENTS

It is important for scientists to think ahead, to attempt to guess what areas are ripe for investigation and which may be bound for oblivion. If one considers recent remarkable leaps in information science, medicine, and particle physics that have materialized in the past century, it is clear that predictions are bound to be less impressive than eventualities, but there are some obvious starting points. Some of those, especially related to the alkali and alkaline earth metals, are suggested here.

NEW PHYSICS

The future of the alkali and alkaline earth metals seems to lie in further investigations of currently intriguing problems rather than radically new phenomena. Those investigations involve questions of astrophysical abundance, Earth's dynamo, time measurement standards, medical diagnosis, and—perhaps most intriguingly—new materials.

New astrophysical knowledge regarding beryllium could lead to answers about the beginning of the universe. The standard theory of the big bang provides for at most 1 percent beryllium production. In some very old stars, however, up to 1,000 times more beryllium than expected has been observed. Some astrophysicists have speculated that a change in the distribution of primordial matter could have resulted from an exotic subatomic phase change during the earliest moments of the universe. Scientists may be able to directly test this hypothesis when the Large Hadronic Collider begins experiments designed to re-create conditions only microseconds after the big bang.

Regarding astrophysical phenomena observable by telescope, the $^6Li/^7Li$ abundance in stars varies widely, and the production mechanisms are far from well understood. Future research is needed to examine the relative roles of convection, diffusion, and spallation for a greater understanding of stellar evolution.

Astrophysical strontium production, too, has shown itself to be of future interest, especially as regards a strontium filament of the exceedingly luminous star Eta Carinae: Its very existence has generated a new way of thinking about supernovae. Supernovae are known to produce large quantities of neutrinos, but also their antiparticles, called antineutrinos. Antineutrinos can change a proton into a neutron, and may play a major role in the building of rare isotopes during stellar explosions. This possibility needs to be explored further.

Closer to home, but no less fascinating, are laboratory experiments that simulate the extremely high pressures of the Earth's interior. These show a new phase of a magnesium silicate ($MgSiO_3$) that bonds in two directions rather than three. Such a phase of matter could indeed flow and help drive convection deep underground, and may, upon further investigation, explain minute changes in Earth's rotation.

Other Earth-based research into laser cooling of cesium may result in higher and higher accuracies of timekeeping standards. Such precision could lead to more rigorous tests of general relativity and help understand variations in pulsars. New alkaline earth–based materials will also continue to be explored and developed.

NEW CHEMISTRY

In all likelihood, nuclear scientists will one day synthesize a few atoms each of elements 119 and 120. Element 119 can be expected to be an alkali metal, and element 120 will be an alkaline earth. However, these elements are predicted to have extremely short half-lives—on the order of just a few microseconds. Unless something surprising happens, it is not to be expected that enough atoms will be synthesized to study the chemical and physical properties of these elements or to find any useful applications for them.

The existing members of the alkali and alkaline earth elements, however, should provide ample opportunity for scientific advancement. Even simple binary materials—like magnesium diboride (MgB_2), which has shown interesting superconducting properties—will most likely bring future revelations. Researchers that synthesize bone for hip and knee replacements are looking for new ceramics that are porous without being fragile. Getting just the right combination of calcium and phosphate compounds seems to be crucial. Ideally, they will formulate a material that bonds well with healthy bone cells and is easily manufactured and machined. Overall, it is to be expected that the fields of medicine, sports, aerospace, and electronics will all benefit from further research into the alkali and alkaline earth elements.

SI Units and Conversions

UNIT	QUANTITY	SYMBOL	CONVERSION
Base units			
meter	length	m	1 m = 3.2808 feet
kilogram	mass	kg	1 kg = 2.205 pounds
second	time	s	
ampere	electric current	A	
kelvin	thermodynamic temperature	K	1 K = 1°C = 1.8°F
candela	luminous intensity		
mole	amount of substance	d mol	
Supplementary Units			
radian	plane angle	rad	π / 2 rad = 90°
steradian	solid angle	sr	
Derived Units			
coulomb	quantity of electricity	C	
cubic meter	volume	m^3	1 m^3 = 1.308 yards^3
farad	capacitance	F	
henry	inductance	H	
hertz	frequency	Hz	
joule	energy	J	1 J = 0.2389 calories
kilogram per cubic meter	density	kg m^{-3}	1 kg m^{-3} = 0.0624 lb. ft^{-3}
lumen	luminous flux	lm	
lux	illuminance	lx	
meter per second	speed	m s^{-1}	1 m s^{-1} = 3.281 ft s^{-1}

UNIT	QUANTITY	SYMBOL	CONVERSION
meter per second squared	acceleration	$m\ s^{-2}$	
mole per cubic meter	concentration	$mol\ m^{-3}$	
newton	force	N	1 N = 7.218 lb. force
ohm	electric resistance	Ω	
pascal	pressure	Pa	$1\ Pa = \dfrac{0.145\ lb}{in^{-2}}$
radian per second	angular velocity	$rad\ s^{-1}$	
radian per second squared	angular acceleration	$rad\ s^{-2}$	
square meter	area	m^2	$1\ m^2 = 1.196\ yards^2$
tesla	magnetic flux density	T	
volt	electromotive force	V	
watt	power	W	$1W = 3.412\ Btu\ h^{-1}$
weber	magnetic flux	Wb	

PREFIXES USED WITH SI UNITS		
PREFIX	**SYMBOL**	**VALUE**
atto	a	$\times 10^{-18}$
femto	f	$\times 10^{-15}$
pico	p	$\times 10^{-12}$
nano	n	$\times 10^{-9}$
micro	μ	$\times 10^{-6}$
milli	m	$\times 10^{-3}$
centi	c	$\times 10^{-2}$
deci	d	$\times 10^{-1}$
deca	da	$\times 10$
hecto	h	$\times 10^{2}$
kilo	k	$\times 10^{3}$
mega	M	$\times 10^{6}$
giga	G	$\times 10^{9}$
tera	T	$\times 10^{12}$

List of Acronyms

AGB   asymptotic giant branch

DNA   deoxyribonucleic acid

EDTA   ethylenediaminetetraacetic acid

FDA   Food and Drug Administration

ISM   interstellar medium

RNA   ribonucleic acid

SneI   Type-1 supernovae

TSP   trisodium phosphate

Periodic Table
of the Elements

Periodic Table of the Elements

Atomic number
Symbol
Atomic weight

3
Li
6.941

Legend: Halogens, Metals, Nonmetals, Metalloids, Unknown

Numbers in parentheses are atomic mass numbers of most stable isotopes.

1 IA	2 IIA	3 IIIB	4 IVB	5 VB	6 VIB	7 VIIB	8 VIIIB	9 VIIIB	10 VIIIB	11 IB	12 IIB	13 IIIA	14 IVA	15 VA	16 VIA	17 VIIA	18 VIIIA
1 H 1.00794																	2 He 4.0026
3 Li 6.941	4 Be 9.0122											5 B 10.81	6 C 12.011	7 N 14.0067	8 O 15.9994	9 F 18.9984	10 Ne 20.1798
11 Na 22.9898	12 Mg 24.3051											13 Al 26.9815	14 Si 28.0855	15 P 30.9738	16 S 32.067	17 Cl 35.4528	18 Ar 39.948
19 K 39.0938	20 Ca 40.078	21 Sc 44.9559	22 Ti 47.867	23 V 50.9415	24 Cr 51.9962	25 Mn 54.938	26 Fe 55.845	27 Co 58.9332	28 Ni 58.6934	29 Cu 63.546	30 Zn 65.409	31 Ga 69.723	32 Ge 72.61	33 As 74.9216	34 Se 78.96	35 Br 79.904	36 Kr 83.798
37 Rb 85.4678	38 Sr 87.62	39 Y 88.906	40 Zr 91.224	41 Nb 92.9064	42 Mo 95.94	43 Tc (98)	44 Ru 101.07	45 Rh 102.9055	46 Pd 106.42	47 Ag 107.8682	48 Cd 112.412	49 In 114.818	50 Sn 118.711	51 Sb 121.760	52 Te 127.60	53 I 126.9045	54 Xe 131.29
55 Cs 132.9054	56 Ba 137.328	57-70 ☆	72 Hf 178.49	73 Ta 180.948	74 W 183.84	75 Re 186.207	76 Os 190.23	77 Ir 192.217	78 Pt 195.08	79 Au 196.9655	80 Hg 200.59	81 Tl 204.3833	82 Pb 207.2	83 Bi 208.9804	84 Po (209)	85 At (210)	86 Rn (222)
87 Fr (223)	88 Ra (226)	89-102 ★	104 Rf (261)	105 Db (262)	106 Sg (266)	107 Bh (262)	108 Hs (263)	109 Mt (268)	110 Ds (271)	111 Rg (272)	112 Cn (277)	113 Uut (284)	114 Uuq (285)	115 Uup (288)	116 Uuh (292)	117 Uus	118 Uuo (294)

☆ Lanthanides

57 La 138.9055	58 Ce 140.115	59 Pr 140.908	60 Nd 144.24	61 Pm (145)	62 Sm 150.36	63 Eu 151.966	64 Gd 157.25	65 Tb 158.9253	66 Dy 162.500	67 Ho 164.9303	68 Er 167.26	69 Tm 168.9342	70 Yb 173.04	71 Lu 174.967

★ Actinides

89 Ac (227)	90 Th 232.0381	91 Pa 231.036	92 U 238.0289	93 Np (237)	94 Pu (244)	95 Am 243	96 Cm (247)	97 Bk (247)	98 Cf (251)	99 Es (252)	100 Fm (257)	101 Md (258)	102 No (259)	103 Lr (260)

© Infobase Publishing

124

Table of
Element Categories

Element Categories

Nonmetals
1	H	Hydrogen
6	C	Carbon
7	N	Nitrogen
8	O	Oxygen
15	P	Phosphorus
16	S	Sulfur
34	Se	Selenium

Halogens
9	F	Fluorine
17	Cl	Chlorine
35	Br	Bromine
53	I	Iodine
85	At	Astatine

Noble Gases
2	He	Helium
10	Ne	Neon
18	Ar	Argon
36	Kr	Krypton
54	Xe	Xenon
86	Ra	Radon

Metalloids
5	B	Boron
14	Si	Silicon
32	Ge	Germanium
33	As	Arsenic
51	Sb	Antimony
52	Te	Tellurium
84	Po	Polonium

Alkali Metals
3	Li	Lithium
11	Na	Sodium
19	K	Potassium
37	Rb	Rubidium
55	Cs	Cesium
87	Fr	Francium

Alkaline Earth Metals
4	Be	Beryllium
12	Mg	Magnesium
20	Ca	Calcium
38	Sr	Strontium
56	Ba	Barium
88	Ra	Radium

Post-Transition Metals
13	Al	Aluminum
31	Ga	Gallium
49	In	Indium
50	Sn	Tin
81	Tl	Thallium
82	Pb	Lead
83	Bi	Bismuth

Transactinides
104	Rf	Rutherfordium
105	Db	Dubnium
106	Sg	Seaborgium
107	Bh	Bohrium
108	Hs	Hassium
109	Mt	Meitnerium
110	Ds	Darmstadtium
111	Rg	Roentgenium
112	Cn	Copernicium
113	Uut	Ununtrium
114	Uuq	Ununquadium
115	Uup	Ununpentium
116	Uuh	Ununhexium
118	Uuo	Ununoctium

Transition Metals
21	Sc	Scandium	39	Y	Yttrium	72	Hf	Hafnium
22	Ti	Titanium	40	Zr	Zirconium	73	Ta	Tantalum
23	V	Vanadium	41	Nb	Niobium	74	W	Tungsten
24	Cr	Chromium	42	Mo	Molybdenum	75	Re	Rhenium
25	Mn	Manganese	43	Tc	Technetium	76	Os	Osmium
26	Fe	Iron	44	Ru	Ruthenium	77	Ir	Iridium
27	Co	Cobalt	45	Rh	Rhodium	78	Pt	Platinum
28	Ni	Nickel	46	Pd	Palladium	79	Au	Gold
29	Cu	Copper	47	Ag	Silver	80	Hg	Mercury
30	Zn	Zinc	48	Cd	Cadmium			

Note: The organization of periodic table of the elements, while useful to chemists and physicists, may be confusing to nonscientists in that some groupings of similar elements appear as vertical columns (halogens, for example), some as horizontal rows (lanthanides, for example), and some as a combination of both (nonmetals).

Lanthanides
57	La	Lanthanum	62	Sm	Samarium	67	Ho	Holmium
58	Ce	Cerium	63	Eu	Europium	68	Er	Erbium
59	Pr	Praseodymium	64	Gd	Gadolinium	69	Tm	Thulium
60	Nd	Neodymium	65	Tb	Terbium	70	Yb	Ytterbium
61	Pm	Promethium	66	Dy	Dysprosium	71	Lu	Lutetium

The table of element categories is intended as a quick reference sheet to easily determine which elements belong to which groups. (Element 117 does not appear in this list because it is undiscovered as of the publishing of this book.)

Actinides
89	Ac	Actinium	94	Pu	Plutonium	99	Es	Einsteinium
90	Th	Thorium	95	Am	Americium	100	Fm	Fermium
91	Pa	Protactinium	96	Cm	Curium	101	Md	Mendelevium
92	U	Uranium	97	Bk	Berkelium	102	No	Nobelium
93	Np	Neptunium	98	Cf	Californium	103	Lr	Lawrencium

Chronology

1743 French mineralogist René-Just Haüy is born on February 28 in Saint-Just-en-Chaussée, France.

1778 English chemist Sir Humphrey Davy is born on December 17 in Penzance, England.

1779 Swedish chemist Jöns Jakob Berzelius is born on August 20 in Väversunda, Sweden.

1787 German optician Joseph von Fraunhofer is born on March 6 in Straubing, Bavaria.

1792 Swedish chemist Johan August Arfwedson is born on January 12

1794 French chemist Antoine-Alexandre-Brutus Bussy is born on May 29 in Marseille, France.

1798 René-Just Haüy shows that beryl and emeralds have identical compositions.

1800 German chemist Friedrich Wöhler is born on July 31 in Eschersheim, Germany.

1807 Humphrey Davy discovers sodium and potassium.

1808 Humphrey Davy isolates strontium, barium, calcium, and magnesium.

John Dalton publishes *A New System of Chemical Philosophy*.

1811 German chemist Robert Wilhelm Bunsen is born on March 31 in Göttingen, Germany.

1814 Joseph von Fraunhofer invents the spectroscope and discovers numerous dark lines in the Sun's spectrum, which he is unable to explain, but which are now called *Fraunhofer lines*.

1817 Jöns Jakob Berzelius isolates selenium.

1818 Johan Arfwedson identifies lithium as a new element.

1822 René-Just Haüy dies on June 3 in Paris, France.

1824   German physicist Gustav Robert Kirchhoff is born on March 12 in Königsberg, Prussia.

1826   Joseph von Fraunhofer dies on June 7 in Munich, Germany.

1828   Friedrich Wöhler and Antoine Bussy isolate beryllium.

1829   Sir Humphrey Davy dies on May 29 in Geneva, Switzerland.

1834   Russian chemist Dmitri Mendeleev is born on February 8 in Tobolsk, Siberia.

1841   Johan August Arfwedson dies on October 28.

1844   John Dalton dies on July 27 in Manchester, England.

1848   Jöns Jakob Berzelius dies on August 7 in Stockholm, Sweden.

1856   English physicist J. J. Thomson is born on December 18 in Cheethan Hill, Manchester, England.

1858   German physicist Max Karl Ernst Ludwig Planck is born on April 23 in Kiel, Holstein, Germany.

1859   French physicist Pierre Curie is born on May 15 in Paris, France.

1860   Robert Bunsen and Gustav Kirchhoff discover cesium.

1861   Robert Bunsen and Gustav Kirchhoff discover rubidium.

1867   French-Polish chemist Marie Curie is born November 7 in Warsaw, Poland.

1870   French physicist Jean-Baptiste Perrin is born on September 30 in Lille, France.

1871   British physicist Ernest Rutherford is born on August 30 in Brightwater, New Zealand.

1872   French physicist Paul Langevin is born on January 23 in Paris, France.

        German chemist Richard Willstätter is born on August 13 in Karlsruhe, Baden, Germany.

1878   German-Austrian physicist Lise Meitner is born on November 17 in Vienna, Austria.

1879 German chemist Otto Hahn is born on March 8 in Frankfurt am Main, Germany.

1882 Friedrich Wöhler dies on September 23 in Göttingen, Germany.

Antoine Bussy dies on March 1 in Paris, France.

1887 Austrian physicist Erwin Schrödinger is born on August 12 in Vienna, Austria.

Gustav Kirchhoff dies on October 17 in Berlin, Germany.

1892 French physicist Louis-Victor-Pierre-Raymond, 7th duc de Broglie, is born on August 15 in Dieppe, France.

1897 J. J. Thomson reports his discovery of the electron.

1898 Pure beryllium is prepared electrolytically.

Pierre and Marie Curie discover polonium and radium.

1899 Robert Bunsen dies on August 16 in Heidelberg, Germany.

1900 Max Planck develops the equation that describes blackbody radiation by introducing the concept of quantization of energy.

1901 Italian physicist Enrico Fermi is born on September 29 in Rome, Italy.

1902 German chemist Fritz Strassmann is born on February 22 in Boppard, Germany.

1903 Pierre and Marie Curie share the Nobel Prize in physics with French physicist Henri Becquerel for their work in radiation research.

1905 Richard Willstätter begins the work that leads to the determination of the structure of chlorophyll.

1906 Pierre Curie dies on April 19 in Paris, France.

J. J. Thomson receives the Nobel Prize in physics "in recognition of the great merits of his theoretical and experimental investigations on the conduction of electricity by gases."

1907 Dmitri Mendeleev dies on January 20 in St. Petersburg, Russia.

1908 J. J. Thomson is knighted by King Edward VII of England.

Ernest Rutherford receives the Nobel Prize in chemistry "for his investigations into the disintegration of the elements, and the chemistry of radioactive substances."

1909 French physicist Marguerite Perey is born on October 19 in Villemomble, France.

1911 Ernest Rutherford announces the discovery of the atomic nucleus.

Marie Curie receives the Nobel Prize in chemistry for the discovery of radium.

1914 Ernest Rutherford is knighted by King George V of England.

1915 Richard Willstätter receives the Nobel Prize in chemistry for his work on photosynthesis and the structure of chlorophyll.

1917 Sir Ernest Rutherford achieves the first transmutation of an element, converting nitrogen into oxygen.

1918 Max Planck receives the Nobel Prize in physics for "his discovery of energy quanta."

1924 Louis de Broglie introduces the theory of the wave nature of matter.

1926 Erwin Schrödinger publishes the paper that introduces the Schrödinger equation to explain the electronic structure of the hydrogen atom.

Jean-Baptiste Perrin receives the Nobel Prize in physics in part for his work on the study of the structure of matter.

1928 Japanese-American chemist Osamu Shimomura is born on August 27 in Kyoto, Japan.

1929 Louis de Broglie receives the Nobel Prize in physics "for his discovery of the wave nature of electrons."

1933 Erwin Schrödinger and Paul Dirac share the Nobel Prize in physics "for the discovery of new productive forms of atomic theory."

1934 Marie Curie dies on July 4 in Sancellemox, Savoy, France.

1937   Sir Ernest Rutherford dies on October 19 in Cambridge, England.

1938   Nuclear fission is discovered by Otto Hahn, Fritz Strassmann, and Lise Meitner.

Enrico Fermi receives the Nobel Prize in physics for "the discovery of nuclear reactions brought about by slow neutrons."

1939   Marguerite Perey isolates francium.

1940   Sir J. J. Thomson dies on August 30 in Cambridge, England.

1942   Jean-Baptiste Perrin dies on April 17 in New York, New York.

Richard Willstätter dies on August 2 in Muralto, Locarno, Switzerland.

1944   Otto Hahn receives the Nobel Prize in chemistry for the discovery of nuclear fission.

1946   Paul Langevin dies on December 19 in Paris, France.

1947   American chemist Martin Chalfie is born on January 15 in Chicago, Illinois.

Max Planck dies on October 4 in Göttingen, West Germany.

1952   American chemist Roger Y. Tsien is born on February 1 in New York, New York.

1954   Enrico Fermi dies on November 28 in Chicago, Illinois.

1961   Erwin Schrödinger dies on January 4 in Vienna, Austria.

1962   Osamu Shimomura performs the research on the green fluorescent protein that leads to the Nobel Prize in chemistry.

1968   Otto Hahn dies on July 28 in Göttingen, Germany.

Lise Meitner dies on October 27 in Cambridge, England.

1975   Marguerite Perey dies on May 13 in Louveciennes, France.

1980   Fritz Strassmann dies on April 22 in Mainz, Germany.

1987   Louis de Broglie dies on March 19 in Louveciennes, France.

1992   Martin Chalfie and Roger Y. Tsien independently perform the research that leads to the Nobel Prize in chemistry.

1997   The International Union for Pure and Applied Chemistry affirms naming element 109 "meitnerium" in honor of Lise Meitner.

Nobel Prize awarded to Steven Chu, Claude Cohen-Tannoudji, and William Phillips for development of methods to cool and trap atoms with laser light.

2007   Former U.S. vice president Al Gore shares the Nobel Peace Prize.

2008   Osamu Shimomura, Martin Chalfie, and Roger Y. Tsien each receives 1/3 of the Nobel Prize in chemistry for the discovery and development of the green fluorescent protein.

2009   U.S. president Barack Obama selects Nobel Prize–winning physicist Steven Chu as energy secretary.

Glossary

accelerator an apparatus that increases the kinetic energy of charged particles.

acid a type of compound that contains hydrogen and dissociates in water to produce hydrogen ions.

actinide the elements ranging from thorium (atomic number 90) to lawrencium (number 103); they all have two outer electrons in the 7s subshell plus increasingly more electrons in the 5f subshell.

active metal a metal that is easily oxidized; an active metal reacts readily with other elements.

alkali metal the elements in column IA of the periodic table (exclusive of hydrogen); they all are characterized by a single valence electron in an s subshell.

alkaline earth metal the elements in column IIA of the periodic table; they all are characterized by two valence electrons that fill an s subshell.

alkyl group a group of carbon and hydrogen atoms such as methyl ($-CH_3$), ethyl ($-CH_2CH_3$), propyl ($-CH_2CH_2CH_3$), and so forth.

alloy a material consisting of two or more metals, or a metal and a nonmetal.

alpha decay a mode of radioactive decay in which an alpha particle—a nucleus of helium 4—is emitted; the daughter isotope has an atomic number two units less than the atomic number of the parent isotope, and a mass number that is four units less.

amalgam an alloy of mercury with one or more other metals.

amphoteric describing a compound that reacts with both acids and bases.

angular momentum referring to an object moving along a curved path, the mathematical product of the object's mass and velocity and the radius of curvature of the path.

anhydrous a chemical compound that lacks any water of crystallization.

anion an atom with one or more extra electrons giving it a net negative charge.

anode the electrode in an electrochemical circuit at which an oxidation half-reaction takes place.

antiseptic any substance that is nontoxic to the body but that kills disease-causing microorganisms.

aquatic describing an environment in water such as a pond, river, lake, or ocean.

aqueous describing a solution in water.

arteriole a blood vessel that leads to capillaries from an artery.

assay to determine the chemical composition of an ore or mineral.

asymptotic giant branch the area of the Hertzsprung-Russell diagram above and to the left of the red giant branch.

atmosphere the mixture of gases surrounding Earth; *atmos* = "vapor."

atom the smallest part of an element that retains the element's chemical properties; atoms consist of protons, neutrons, and electrons.

atomic mass (also called atomic weight) the total mass (or weight) of all the components of an atom, often cited relative to the mass of carbon 12.

atomic number the number of protons in an atom of an element; the atomic number establishes the identify of an element.

base a substance that reacts with an acid to give water and a salt; a substance that, when dissolved in water, produces hydroxide ions.

berylliosis a lung disease associated with exposure to beryllium.

beta decay a mode of radioactive decay in which a beta particle—an ordinary electron—is emitted; the daughter isotope has an atomic number one unit greater than the atomic number of the parent isotope, but the same mass number.

bicarbonate of soda sodium bicarbonate, $NaHCO_3$.

binary compound   a chemical compound consisting of only two elements, but variable numbers of atoms of each of those elements.

bipolar disorder   a psychological condition characterized by large mood swings.

black powder   the traditional form of gunpowder, consisting of charcoal, sulfur, and potassium.

Bose-Einstein condensate   a state of matter, produced in laboratories, in which atoms are packed so close together that their wavefunctions become correlated similar to those of photons in a laser beam, and coherent matter waves can be formed.

brachytherapy   injection of a small source or "seed" to a specific physiological site.

calcareous   composed of or containing calcium carbonate.

calibration curve   a plot constructed from known data with which scientists can compare a known value of a variable to get the corresponding value of the related variable.

capacitor   a device that stores electrical charge.

carbonate   the ion CO_3^{2-} or a compound that contains that ion.

catalyst   a material substance that increases the rate at which a chemical reaction takes place without itself undergoing any chemical change.

cathode   the electrode in an electrochemical circuit at which a reduction half-reaction takes place.

cathodic protection   the process in which the oxidation of a more active metal, e.g., zinc or magnesium, protects an iron or steel structure; the more active metal is said to have been "sacrificed" in order to protect the underlying structure.

cation   an atom that has lost one or more electrons to acquire a net positive charge.

caustic   a description of a substance or solution that is strongly alkaline.

caustic potash   potassium hydroxide (KOH).

caustic soda   sodium hydroxide (NaOH).

chalk   a powdery form of calcium carbonate, $CaCO_3$.

chelate   an ion or molecule that attaches to a metal ion at two or more positions.

chemical bond   a strong electrostatic attraction between atoms within a molecule (*see* **covalent bond**) or between two ions in crystalline solids (*see* **ionic bond**).

chemical change   a change in which one or more chemical elements or compounds form new compounds; in a chemical change, the names of the compounds change.

chlorophyll   the pigment responsible for the green color of most plants.

cobalt-blue glass   a deep blue colored glass made that way by the addition of cobalt.

complexing agent   a chemical compound that bonds at several sites to another ion or molecule.

complex ion   any ion that contains more than one atom.

compound   a pure chemical substance consisting of two or more elements in fixed, or definite, proportions.

convection   molecular movement within gases or fluids that effects heat exchange.

covalent bond   a chemical bond formed by sharing valence electrons between two atoms (in contrast to an ionic bond, in which one or more valence electrons are transferred from one atom to another atom).

critical current density   typically referring to superconductors, the upper limit to the number of electrons per second passing through a cross-sectional area of a wire at which the superconducting properties of the material can be maintained.

critical point temperature   the temperature of a pure substance at which the liquid and gaseous phases become indistinguishable.

criticality   in weapons or reactor physics, the point at which the mass of fuel is sufficient to sustain continued nuclear fission.

crucible   a ceramic or metallic dish in which substances can be heated to high temperatures.

cryogenic   the production of very low temperatures and the study of the properties of materials at low temperatures.

CT scan computerized tomography scan that helps visualize internal organs and other soft tissue.

Curie temperature (or Curie point) the temperature above which a magnetic substance loses its magnetism.

cyclotron a circular particle accelerator that employs the directional component of acceleration rather than the speed component.

dark energy a hypothetical form of energy believed to fill all of space and cause the universe to expand at an accelerated rate.

dark matter a hypothetical form of matter that does not emit radiation but affects visible matter around it gravitationally.

daughter the product of a radioactive decay event; what is formed when a **parent isotope** decays.

decant to pour a solution from one container into another. Most commonly, a supernatant liquid is poured into another container, leaving a precipitate behind.

dessicant a substance that removes moisture from its surroundings.

diffusion the process by which substances mix due to the random motions of their molecules.

distillation a process used to purify liquids or to separate the components of a mixture by successively boiling the liquid, condensing the vapor, and collecting the new liquid or fractions of the original liquid.

diuretic a substance that enhances urine formation by the kidney, often used as hypertension medication.

double salt a salt that contains two metal ions.

Earth's crust the hard surface layer, which is about 10 kilometers (6.2 miles) thick under the oceans and about three times that thickness on the continents.

Earth's mantle the moving, flowing layer beneath the crust.

electrical conductivity the ability of a substance, such as a metal or a solution, to conduct an electrical current.

electrochemistry the study of chemical reactions that take place in galvanic or electrolytic cells.

electrolysis   the process by which a chemical reaction is made to take place by passing an electrical current through a chemical solution.

electron   a subatomic particle found in all neutral atoms; possesses the negative charges in atoms.

electron affinity   the energy that is released when an electron is added to an isolated atom or ion; a measure of how strongly that electron is attracted to the atom or ion.

electronic configuration   a description of the arrangement of the electrons in an atom or ion, showing the number of electrons occupying each subshell.

electropositive   a description of elements, usually metals, that tend to lose electrons and form positive ions.

electrostatic   the type of interaction that exists between electrically charged particles; electrostatic forces attract particles together if the particles have charges of opposite sign, while the forces cause particles that have charges of like sign to repel each other.

element   a pure chemical substance that contains only one kind of atom.

enzyme   a type of protein molecule that acts as a catalyst in biochemical reactions.

Epsom salt   a laxative or soaking agent containing $MgSO_4 \times 7H_2O$.

evaporation   the phase change in which a liquid is converted into its vapor.

exothermic   describes any chemical reaction that releases heat to its surroundings.

fallout   radioactive particles deposited from the atmosphere from either a nuclear explosion or a nuclear accident.

family   *see* **group.**

fermentation   a chemical reaction due to microbial activity; a familiar example is the conversion of sugar into alcohol.

fission   *see* **nuclear fission.**

fixing agent   a chemical compound that removes unexposed silver in the development of a photographic image.

fluorescence   the spontaneous emission of light from atoms or molecules when electrons make transitions from states of higher energy to states of lower energy.

flux   in physics, the rate of flow of anything through a given surface area. In metallurgy, a substance that promotes the fusion of two metals or minerals.

Fraunhofer lines   absorption lines in the Sun's spectrum discovered by Joseph von Fraunhofer.

fuel cell   a cell in which the chemical energy stored in the fuel is converted directly into useful electrical energy.

fumigant   a substance that releases fumes to exterminate insects.

fungicide   a pesticide that targets fungi such as mold and mildew.

fusion   *see* **nuclear fusion.**

galvanic action   the spontaneous occurrence of an electrochemical reaction.

gamma decay   a mode of radioactive decay in which a very-high-energy photon of electromagnetic radiation—a gamma ray—is emitted; the daughter isotope has the same atomic number and mass number as the parent isotope, but lower energy.

gastrointestinal tract   the gastrointestinal tract begins with the mouth, leads to the esophagus, extends through the stomach, small and large intestines, and ends at the anus.

general relativity   the theory developed by Albert Einstein that explains gravity in terms of a four-dimensional space-time universe.

Glauber's salt   hydrated sodium sulfate ($Na_2SO_4 \times 10H_2O$).

global positioning system (GPS)   a system of artificial satellites that are used worldwide for navigation, mapmaking, land surveying, tracking, and surveillance.

goiter   an enlargement of the thyroid gland, usually apparent as a swelling of the neck.

group (or family)   the elements that are located in the same column of the periodic table; also called a family, elements in the same column have similar chemical and physical properties.

half-life   the time required for half of the original nuclei in a sample to decay; during each half-life, half of the nuclei that were present at the beginning of that period will decay.

halide ion   a simple ion with a charge of −1 formed by adding an electron to a neutral halogen atom.

halogen   the elements in column VIIB of the periodic table; all of them share a common set of seven valence electrons in an nth energy level such that their outermost electronic configuration is ns^2np^5.

heat of hydration   the energy change that occurs when an ion becomes surrounded by water molecules.

hemoglobin   the iron-carrying protein that makes blood red.

Hertzsprung-Russell diagram   used in astrophysics, a graph that plots luminosity versus surface temperature of a star.

homogeneous   having the same distribution or density in every direction.

Homo habilis   an ancient cousin of *Homo sapiens* that lived about 1.75 million years ago.

hydrate   a crystalline substance in which water molecules have combined with an ionic compound.

hydride   the ion H^- or a compound that contains that ion.

hydrocarbon   chemical compounds that contain only the elements carbon and hydrogen.

hydrosphere   the water that is contained in Earth's oceans, lakes, rivers, and other surface bodies of water; *hydro* = "water."

hydroxide   the ion OH^- or a compound that contains that ion.

hypertension   high blood pressure.

hypo   sodium thiosulfate ($Na_2S_2O_3$); also called sodium hyposulfite.

hypothalamus   an area of the brain that helps the body regulate hormone and fluid flow as well as sleep and appetite.

incendiary   describing a chemical reaction that generates sufficient heat to start a fire.

inert   an element that has little or no tendency to form chemical bonds; the inert gases are also called **noble gases.**

interstellar medium (ISM)   the regions of galaxies between stars.

ion   an atom or group of atoms that has a net electrical charge.

ionic bond   a strong electrostatic attraction between a positive ion and a negative ion that holds the two ions together.

isotope   a form of an element characterized by a specific mass number; the different isotopes of an element have the same number of protons but different numbers of neutrons, hence different mass numbers.

lanthanide   the elements ranging from cerium (atomic number 58) to lutetium (number 71); they all have two outer electrons in the 6s subshell plus increasingly more electrons in the 4f subshell.

lime   a chemical consisting of calcium oxide, CaO.

lithosphere   the solid part of Earth; its crust, mantle, and core; *litho* = "stone."

luminescence   the emission of light by an object for any reason besides an increase in temperature.

luminosity   apparent brightness, especially as pertains to stars; technically, the energy emitted per second.

luster   the shininess associated with the surfaces of most metals.

magma   hot molten material that originates in **Earth's crust** or **mantle.**

magnesia   magnesium oxide (MgO).

main group element   an element in one of the first two columns or one of the right-hand six columns of the periodic table; distinguished from transition metals, which are located in the middle of the table, and from rare earths, which are located in the lower two rows shown apart from the rest of the table.

main sequence   the area of the Hertzsprung-Russell diagram where most stars tend to be located during their early evolution.

mass   a measure of an object's resistance to acceleration; determined by the sum of the elementary particles comprising the object.

mass number   the sum of the number of protons and neutrons in the nucleus of an atom. *See also* **isotope.**

metal any of the elements characterized by being good conductors of electricity and heat in the solid state; approximately 75 percent of the elements are metals.

metalloid (also called semi-metal) any of the elements intermediate in properties between the metals and nonmetals; the elements in the periodic table located between metals and nonmetals.

metallurgy the technology of producing metals from their ores, purifying the metals, and fashioning alloys from them.

microwave long-wavelength (millimeters to a few centimeters), low-frequency electromagnetic radiation invisible to the human eye.

mixture a combination of two or more substances in which the substances retain their original properties (as opposed to a **compound** that likely has properties completely distinct from its original components).

moderator in a nuclear reactor, a medium that slows the speed of initially fast-moving neutrons.

molten a liquid state usually characterized by temperatures much higher than 77°F (25°C).

native metal a metal found in nature in its pure state; familiar examples are gold, silver, and copper.

neutron the electrically neutral particle found in the nuclei of atoms.

nichrome referring to an alloy that contains mostly nickel, some chromium, and possibly small amounts of other metals.

niter *see* **saltpeter.**

nitrate the ion NO_3^- or a compound that contains that ion.

nitride the ion N^{3-} or a compound that contains that ion.

noble gas any of the elements located in the last column of the periodic table—usually labeled column VIII or 18, or possibly column 0.

nonmetal the elements on the far right-hand side of the periodic table that are characterized by little or no electrical or thermal conductivity, a dull appearance, and brittleness.

nuclear fission the process in which certain isotopes of relatively heavy atoms such as uranium or plutonium break apart into fragments of comparable size; accompanied by the release of large amounts of energy.

nuclear fusion   the process in which certain isotopes of relatively light atoms such as hydrogen or helium can combine to form heavier isotopes; accompanied by the release of large amounts of energy.

nuclear medicine   a branch of medicine that uses radioactive isotopes to detect and treat disease.

nuclear reaction   a process in which atomic nuclei, or nuclei and elementary particles, collide to produce different nuclei.

nucleon   a particle found in the nucleus of atoms; a proton or a neutron.

nucleosynthesis   the process of building up atomic nuclei from protons and neutrons or from smaller nuclei.

nucleus   the small, central core of an atom.

ordinary (chemical) reaction   a process that does not alter the nuclei of the atoms involved. *See also* **chemical change.**

organo-alkali compound   a chemical compound that consists of two parts, one part being an organic group and the other part being an alkali metal.

osmosis   the passage of pure solvent through a semipermeable membrane that separates two solutions having different solute concentrations; a semipermeable membrane is one through which some kinds of molecules pass but other kinds cannot.

osmotic pressure   *see* **osmosis.**

osteoporosis   a bone disease characterized by decreased bone density and an increased risk of bone fracture.

oxidation   an increase in an atom's oxidation state; accomplished by a loss of electrons or an increase in the number of chemical bonds to atoms of other elements. *See also* **oxidation state.**

oxidation potential   the energy per unit charge generated by an oxidation half-reaction compared to the oxidation of hydrogen gas as a reference; the unit of potential is the volt.

oxidation-reduction reaction   a chemical reaction in which one element is oxidized and another element is reduced.

oxidation state   a description of the number of atoms of other elements to which an atom is bonded. A neutral atom or neutral group of atoms of a single element is defined to be in the zero oxidation state. Otherwise, in compounds, an atom is defined as being in a positive or negative oxidation state, depending upon whether the atom is bonded to elements that, respectively, are more or less electronegative than that atom is.

oxide   the ion O^{2-} or a compound that contains that ion.

parent isotope   an atom that undergoes radioactive decay. *See also* **daughter isotope.**

period   any of the rows of the periodic table; rows are referred to as periods because of the periodic, or repetitive, trends in the properties of the elements.

periodic table   an arrangement of the chemical elements into rows and columns such that the elements are in order of increasing atomic number, and elements located in the same column have similar chemical and physical properties.

permanganate   the ion MnO_4^- or a compound that contains that ion.

permeability   the property of a porous material that allows a pressurized gas to pass through it.

peroxide   a compound that contains the group O_2^{2-}; hydrogen peroxide (H_2O_2) is a familiar example.

phosphate   the ion PO_4^{3-} or a compound that contains that ion.

photon   a discrete "particle" with no rest mass or electrical charge that travels at the speed of light.

physical change   any transformation that results in changes in a substance's physical state, color, temperature, dimensions, or other physical properties; the chemical identity of the substance remains unchanged in the process.

physical state   the condition of a chemical substance being either a solid, liquid, or gas.

piezoelectric effect   the production of an electrical current when pressure is applied to a crystal.

planetary nebula   gaseous clouds surrounding a central star that occur when some AGB stars eject part of their atmospheres into the ISM. (Nothing to do with planets, but were once thought to have some such association.)

plankton   microscopic plants or animals that float in aquatic environments.

plasticity   the malleability or pliability of a material.

plate tectonics   the theory that Earth's crust is made of plates that have moved during large spans of geologic time.

polarization   the direction of the electric field of an electromagnetic wave.

porosity   the ability of a material containing pores to allow the passage of gases or liquids through it.

porphyrin   an organic compound that is found in chlorophyll and hemoglobin.

positron emission tomography (PET)   a three-dimensional imaging technique used in nuclear medicine; gamma radiation that accompanies positron (positively charged electron) emission produces the images.

post-transition metal   the metals that occur in positions to the right of the transition metals in the periodic table; aluminum, gallium, indium, tin, thallium, lead, bismuth, polonium, and elements 113–118.

potash   potassium carbonate (K_2CO_3).

pozzolan   porous material that reacts with calcium in mortar to make it less permeable to water.

Precambrian   span of geologic time from Earth's formation until the Cambrian era, which began about 540 million years ago.

precipitate   solid particles that result from a chemical reaction taking place in a liquid solution.

product   the compounds that are formed as the result of a chemical reaction.

proton   the positively charged subatomic particle found in the nuclei of atoms.

pulsar   an astronomical source of radio waves that are emitted with a regular frequency.

qualitative analysis   chemical analysis that answers the question: What is in a compound or in a mixture of compounds?

quantitative analysis   chemical analysis that answers the question: How much of an element or compound is in a mixture?

radioactive   *see* **radioactive decay.**

radioactive decay   the disintegration of an atomic nucleus accompanied by the emission of a subatomic particle or gamma ray.

radioisotope   an isotope of an element that is radioactive. *See also* **isotope.**

radioluminescence   the phenomenon in which visible light is produced when ionizing radiation strikes a material; for example, in the past, watch dials were coated with radium to make the dials glow in the dark.

rare earth element   the metallic elements found in the two bottom rows of the periodic table; the chemistry of their ions is determined by electronic configurations with partially filled f subshells. *See also* **lanthanides** and **actinides.**

reactant   the chemical species present at the beginning of a chemical reaction that rearrange atoms to form new species.

reducing agent   a chemical reagent that causes an element in another reagent to be reduced to a lower oxidation state.

reduction   a decrease in an atom's oxidation state; accomplished by a gain of electrons or a decrease in the number of chemical bonds to atoms of other elements. *See also* **oxidation state.**

resonant   a natural frequency of vibration characterized by a large amplitude.

rock salt   *see* **table salt.**

sacrificial anode   *see* **cathodic protection.**

saltpeter   potassium nitrate (KNO_3).

saturated solution   a solution that contains the maximum amount of solute it can hold at a given temperature.

Schrödinger equation   an equation used in atomic physics to describe the wave function of a particle.

sea salt   a soluble salt found in sea water; the salt found in highest quantity is sodium chloride.

sedimentary   a description of rocks that have formed under pressure by the accumulation and solidification of sediments.

semimetal   *see* **metalloid.**

shell   all of the orbitals that have the same value of the principal energy level, n.

silicate   a mineral or compound that contains a metal, silicon, and oxygen.

smelting   the process of separating a metal from its ore by heating the ore to a heat temperature in the presence of a reducing agent.

soda ash   anhydrous sodium carbonate (Na_2CO_3).

soluble   referring to a substance that dissolves in another substance; typically referring to gases, liquids, or solids that dissolve in a liquid like water.

spallation   the breaking apart of a substance.

spectroscope   an instrument that separates light into its component colors for visual observation.

spontaneous fission (SF)   the fission of a heavy nucleus without any interaction with other particles. *See also* **nuclear fission.**

stalactite   a formation consisting of calcium carbonate that hangs from the ceiling of limestone caves.

stalagmite   a formation consisting of calcium carbonate that protrudes upward from the floor of limestone caves.

stellar evolution   the changes in sizes, luminosities, and other properties of stars as they age.

subatomic particle   actual particles (atoms are a composite) that are smaller than atoms, but not limited to atomic constituents. Neutrinos, for example, are subatomic particles.

sublimation   the change of physical state in which a substance goes directly from the solid to the gas without passing through a liquid state.

sublimation point temperature the temperature of a pure substance at which the solid and gaseous phases become indistinguishable. Only carbon and arsenic are capable of sublimation.

sublime *see* **sublimation.**

subshell all of the orbitals of a principal shell that lie at the same energy level.

sulfate the ion SO_4^{2-} or a compound that contains that ion.

supernatant liquid the clear liquid that remains after a precipitate has settled out of solution.

synaptic related to the behavior of synapses, the junction between neurons and other cells that allows them to transmit signals.

syphilis a sexually transmitted disease caused by a bacterial infection; treatable with antibiotics.

table salt sodium chloride (NaCl).

terrestrial referring to Earth or land.

thermal conductivity a measure of the ability of a substance to conduct heat.

thermal neutron a slowly moving neutron; typically used to initiate nuclear fission.

thermonuclear *see* **nuclear fusion.**

transition metal any of the metallic elements found in the 10 middle columns of the periodic table to the right of the alkaline earth metals; the chemistry of their ions largely is determined by electronic configurations with partially filled d subshells.

transmutation the conversion by way of a nuclear reaction of one element into another element; in transmutation, the atomic number of the element must change.

transuranium element any element in the periodic table with an atomic number greater than 92.

triad a grouping of three chemical elements that have similar chemical and physical properties.

triple-alpha process a nuclear fusion reaction in which three helium nuclei combine to produce a carbon nucleus.

tufa a thick rock found in lakes with high concentrations of calcium carbonate.

tumor an abnormal growth of tissue in or on the body.

tweeter a speaker used in audio equipment to reproduce high-frequency sound.

valence a measure of the number of chemical bonds that an atom or ion can form. In the case of a neutral atom, valence is the number of hydrogen atoms to which it can bond. In the case of a simple ion, valence is the charge on the ion.

viscosity the measure of the resistance to flow of a liquid; a viscous liquid tends not to flow.

wavefunction the mathematical description of an atomic or molecular orbital.

white dwarf the leftover core of a high-mass star that has shed its outer layers and now radiates only in the infrared.

Further Resources

The following is a list of sources offering readings related to individual alkali and alkaline earth metal elements.

LITHIUM
Books and Articles

Barondes, Samuel H. *Molecules and Mental Illness.* New York: Scientific American Library, 1993. This book describes various psychological disorders, what causes them, and possible treatments, including the use of lithium to treat bipolar disorder.

Bosfeld, Jane. "Lithium May Be the Answer for Lou Gehrig's Disease." *Discover,* January, 2009. Number 28 of *Discover* magazine's top 100 stories of 2008, this article explains how a new study brings some hope that Lou Gehrig's disease can be treated.

Schou, Mogens. *Lithium Treatment of Mood Disorders: A Practical Guide.* Basel, Switzerland: Karger, 2004. Written by a Danish psychiatrist and professor of biological psychiatry with a strong background in the subject, this guide gives details and answers questions about lithium treatment in language that is easy to understand.

Internet Resources

Delaney, K. "Naval Sea Systems Command Issues Submarines Life-Saving Lithium Hydroxide Curtains Developed by Battelle." Available online. URL: www.battelle.org/SPOTLIGHT/news_archives/archive_04/4-06-04LithCurtain.aspx. Accessed on June 23, 2009. This article describes how lithium hydroxide curtains can remove hazardous carbon dioxide from the atmosphere of a disabled submarine, improving the crewmembers' ability to survive while awaiting rescue.

Encyclopedia Britannica. "Fusion Reactor." Available online. URL: www.britannica.com/EBchecked/topic/222821/fusion-reactor. Accessed on June 23, 2009. A good description of the science of fusion reactors, including lithium's role.

National Alliance on Mental Illness. Available online. URL: http://www.nami.org. Accessed on June 23, 2009. This is the official Web site of the National Alliance on Mental Illness, with information about bipolar disorder, other forms of depression, and their treatments.

Pilcher, Helen R. "The Ups and Downs of Lithium." Available online. URL: www.bioedonline.org/news/news.cfm?art=552. Accessed on June 23, 2009. Article explains the pros and cons of lithium use for manic depression and other brain disorders.

SODIUM

Books and Articles

Goldman, Erik L. "New DASH Findings Push for Lower Sodium Intake." *Family Practice News* 30 (2000): 13. International Medical News Group. This article describes why some hypertension experts recommend minimizing sodium in food.

Greeley, Alexandra. "A Pinch of Controversy Shakes Up Dietary Salt." *FDA Consumer* 31 (November–December 1997): 24. Persistent high blood pressure is one of the most common health conditions, but the salt connection is not a recent discovery, as explained in this article.

MacGregor, G. A., and H. E. de Wardener. *Salt, Diet and Health: Neptune's Poisoned Chalice: The Origins of High Blood Pressure.* Cambridge, U.K.: University Press, 1998. This book outlines the dangers of excessive salt intake to human health as well as the socioeconomic history of sodium chloride.

Internet Resources

Oak Ridge Associated Universities. "Low Sodium Salt Substitutes." Available online. URL: www.orau.org/PTP/collection/consumer%20products/lowsodiumsalt.htm. Accessed on June 23, 2009. This Web site explains the difference between ordinary table salt and salt substitute.

Rohack, J. James. "AMA to *New York Times:* Salt in Our Diet: How Much Is Right?" Available online. URL: www.ama-assn.org/ama/pub/category/16461.html. Accessed on June 23, 2009. This letter from the president-elect of the American Medical Association, published February 13, 2009, recommends Americans reduce dietary salt.

POTASSIUM
Books and Articles

Brown, Michael. "Potassium in Europa's Atmosphere." *Icarus* 151 (2001): 190–195. This article discusses the research that led to a better understanding of the sodium-potassium content of Jupiter's moon Europa.

Dajer, Tony. "Potassium Paralysis." *Discover* 20 (December 1999): 49. Periodic paralysis can be caused by either high or low cellular potassium, as described in this article.

———. "Vital Signs: Potassium Overload." *Discover* 28 (March 2007): 20. This article gives an account of a case of dangerous potassium overdose from eating a durian fruit.

Kamen, Betty. *Everything You Always Wanted to Know about Potassium but Were Too Tired to Ask.* Novato, Calif.: Nutrition Encounter, 1992. Convincing and easy-to-read, this book lays out the need and methods to get more potassium from foods.

Lehrman, Sally. "Sobering Shift." *Scientific American* 290 (April 2004): 22. This article describes how a particular gene controls a gateway for the flow of potassium ions in cells, which could determine sensitivity to alcohol as well as provide a mechanism for intoxication.

Seaborg, Glenn T., and Eric Seaborg. *Adventures in the Atomic Age: From Watts to Washington.* New York: Farrar, Straus, and Giroux, 2001. In a section of his autobiography, Glenn Seaborg relates the story of his discovery of iodine 131 and the isotope's widespread use in nuclear medicine.

Svitil, Kathy A. "A Strange Brew in Middle Earth." *Discover* 24 (August 2003): 16. The potential role of radioactive potassium in the continued heat of the Earth's core, as indicated by the research of geochemist V. Rama Murthy and colleagues, is explored in this article.

Yellen, G. "The Voltage Gated Potassium Channels and Their Relatives." *Nature* 419 (2002): 35. Voltage-gated potassium channels are membrane-signaling proteins that pass millions of ions per second across the membrane with astonishing selectivity. The architectural modules and functional components of these molecules are described here.

Internet Resources

Anderson, J., L. Young, and E. Long. "Potassium and Health." Available online. URL: www.ext.colostate.edu/pubs/foodnut/09355.html. Accessed on June 23, 2009. An easy-to-read resource on potassium and food sources from the Colorado State University Cooperative Extension.

Argonne National Laboratory. Available online. URL: www.ead.anl. gov/pub/doc/potassium.pdf. Accessed on June 23, 2009. Human health fact sheet contains details on potassium sources, uses, and human health risks and benefits.

International Plant Nutrition Institute. Available online. URL: www. ipni.net/ppiweb/canadaw.nsf/$webindex/3044476F390951EB06256F 4000690213?open document&navigator=home+page. Accessed on June 23, 2009. On this Web site, the International Plant Nutrition Institute describes the use of potassium-based fertilizers.

Sanders, Robert. "Radioactive Potassium May Be Major Heat Source in Earth's Core." Available online. URL: berkeley.edu/ news/media/releases/2003/12/10_heat.shtml. Accessed on August 5, 2009. This press release from the University of California, Berkeley, describes an experiment that has simulated conditions deep beneath Earth's crust and tested a geodynamo model involving potassium 40.

RUBIDIUM, CESIUM, AND FRANCIUM

Books and Articles

Audoin, Claude, Bernard Guinot, and Jacques Vanier. "The Measurement of Time: Time, Frequency and the Atomic Clock." *Physics Today* 56 (January 2003): 48. This article discusses the history of the atomic clock and how it works.

Suttle, John F. "The Alkali Metals," in *Comprehensive Inorganic Chemistry,* vol. 6. Edited by M. Cannon Sneed and Robert C. Brasted.

Princeton, N.J.: D. Van Nostrand and Company, 1957. The first section of the book presents an overview of how the metals are prepared, their chemical and physical properties, and their alloys.

Internet Resources

Argonne National Laboratory. Available online. URL: www.ead.anl. gov/pub/doc/cesium.pdf. Accessed on June 23, 2009. Human health fact sheet contains details on cesium sources, uses, and human health risks.

Sullivan, Donald B. "How Does One Arrive at the Exact Number of Cycles of Radiation a Cesium-133 Atom Makes in Order to Define One Second?" *Scientific American,* 16 December 2002. Available online. URL: www.scientificamerican.com/article.cfm?id=how-does-one-arrive-at-th. Accessed on June 23, 2009. A brief narrative about how the cesium beam standard for timekeeping came about.

BERYLLIUM

Internet Resources

Argonne National Laboratory. "Beryllium: Human Health Fact Sheet." Available online. URL: www.ead.anl.gov/pub/doc/ beryllium.pdf. Accessed on June 23, 2009. Human health fact sheet contains details on beryllium sources, uses, and human health risks.

Discovery Channel. "Smashing the Universe's Mysteries." Available online. URL: dsc.discovery.com/space/im/universe-lhc-cern-goldfarb.html. Accessed on October 20, 2009. An interview with Steve Goldfarb, a leading scientist working at the world's largest particle accelerator.

Labrador, David. "Study of Lunar Soil Sheds Light on the Sun's Violent Atmosphere." *Scientific American,* 16 October 2001. Available online. URL: www.scientificamerican.com/article.cfm?id=study-of-lunar-soil-sheds. Accessed on June 24, 2009. Scientists analyzed samples of lunar soil obtained by Apollo 17 astronauts for the presence and distribution of beryllium 10, an unstable isotope produced in the solar atmosphere and carried by solar wind.

Occupational Safety and Health Administration. "Safety and Health Topics: Beryllium." Available online. URL: www.osha.gov/SLTC/beryllium/index.html. Accessed on October 20, 2009. This Web site describes some of the safety and health issues associated with the use of beryllium.

WebMD. "Berylliosis." Available online. URL: www.webmd.com/a-to-z-guides/berylliosis. Accessed on June 24, 2009. Details about beryllium-induced lung disease.

MAGNESIUM
Books and Articles

Canfield, Paul C., and George W. Crabtree. "Magnesium Diboride: Better Late than Never." *Physics Today* 56 (2003): 34. Article describes the history and discovery as well as superconducting properties of magnesium diboride.

Hirose, Kei. "Magical Mantle Tour." *Nature* 440 (April 2006): 27.

Raven, Peter H., Ray F. Evert, and Susan E. Eichhorn. *Biology of Plants,* 7th ed. New York: W. H. Freeman and Co., 2005. In the section on the chemistry of photosynthesis, the role of magnesium is explained.

Internet Resources

Graham, Sarah. "Scientists Develop New Method of Manufacturing Superconducting Magnesium Boride." *Scientific American,* 4 September 2002. Available online. URL: www.scientificamerican.com/article.cfm?id=scientists-develop-new-me. Accessed on June 24, 2009. A brief description of MgB_2 manufacture.

Linus Pauling Institute, Oregon State University. "Micronutrient Information Center: Magnesium." Available online. URL: lpi.oregonstate.edu/infocenter/minerals/magnesium/. Accessed on October 20, 2009. The role of magnesium in human health is explained.

Magnesium Elektron Company. "Magnesium for Defence." Available online. URL: www.magnesium-elektron.com/markets-applications.asp?ID=9. Accessed on October 20, 2009. This Web site describes applications of magnesium in the defense industry.

CALCIUM
Books and Articles

Hively, Will. "Worrying about Milk." *Discover* 21 (August 2000): 44. Discusses the pros and surprising cons of milk consumption for the purpose of calcium intake.

Thompson, Robert, and Kathleen Barnes. *The Calcium Lie: What Your Doctor Doesn't Know Could Kill You.* Brevard, N.C.: InTruth-Press, 2008. This book, written by a medical doctor, explains why calcium supplementation may not be a good idea.

Weaver, Connie M., and Robert P. Heaney, eds. *Calcium in Human Health.* Totowa, N.J.: Humana Press, 2006. A collection of essays by experts in all areas of calcium and human physiology.

Internet Resources

Ballantyne, Coco. "Calcium Might Fight Cancer." *Scientific American,* 24 February 2009. Available online. URL: www.scientificamerican. com/blog/60-second-science/post.cfm?id=calcium-might-fight-cancer-2009-02-24. Accessed on June 23, 2009. This article reports on a study that may show an anticorrelation between calcium intake and the probability of developing certain types of cancer.

National Dairy Council. Available online. URL: www.nationaldairy council.org/NationalDairyCouncil/. Accessed on June 23, 2009. Explanation of the importance of calcium in human nutrition, particularly in building strong bones and preventing osteoporosis.

Wenner, Melinda. "Like the Taste of Chalk? You're in Luck—Humans May Be Able to Taste Calcium." *Scientific American,* 20 August 2008. Available online. URL: www.scientificamerican.com/article. cfm?id=osteoporosis-calcium-taste-chalk. Accessed on June 23, 2009. This article explains why many people do not get enough of the nutrient calcium and develop osteoporosis.

STRONTIUM AND BARIUM
Books and Articles

Richards, Michael, et al. "Strontium Isotope Evidence of Neander-thal Mobility at the Site of Lakonis, Greece Using Laser-Ablation PIMMS." *Journal of Archaeological Science* 35 (May 2008): 1,251.

This article details how scientists used strontium isotope dating of tooth enamel to conclude that some Neanderthals traveled widely, which had been debated by other anthropologists.

Zimmer, Carl. "Making Auroras." *Discover* 13 (1992): 24. This article explains how barium gas can help scientists understand the Sun's magnetic field.

Internet Resources

Argonne National Laboratory. "Strontium: Human Health Fact Sheet." Available online. URL: www.ead.anl.gov/pub/doc/ strontium.pdf. Accessed on June 23, 2009. Human health fact sheet contains details on strontium sources, uses, and human health risks.

Bielo, David. "Home on the Reef: A Majority of Young Fish Return to Birthplace." *Scientific American,* 3 May 2007. Available online. URL: www.scientificamerican.com/article.cfm?id=home-on-the-reef-majority-young-fish-return-to-birthplace. Accessed on June 25, 2009. This article explains how barium injected into fish helped scientists learn about the breeding behavior of fish in a coral reef near Papua, New Guinea.

Mayo Clinic. "Barium Enema." Available online. URL: http://www. mayoclinic.com/health/barium-enema/CO00006. Accessed on June 23, 2009. The use of barium enemas as a diagnostic tool is described.

McCook, Allison. "Molecular Sponge for Nuclear Waste." *Scientific American,* 26 February 2001. Available online. URL: /www. scientificamerican.com/article.cfm?id=molecular-sponge-for-nucl. Accessed on June 25, 2009. This article explains how research is leading to ideas on how to absorb and store radioactive strontium 90 from nuclear waste.

RADIUM

Books and Articles

Borzendowski, Janice. *Marie Curie: Mother of Modern Physics.* New York: Sterling, 2009. A modern treatment of the life and work of Marie Curie.

Curie, Eve. *Madame Curie.* New York: Doubleday, Doran, and Company, 1939 (translated by Vincent Sheean). Eve Curie was one of Pierre and Marie Curie's two daughters. Eve's biography of her mother has long been considered the primary source of information about Marie as a person.

Harvie, David I. *Deadly Sunshine: The History and Fatal Legacy of Radium.* Gloucestershire, U.K.: The History Press, 2005. This book provides a detailed summary of radium's place in human history, from its discovery by Marie Curie to its uses and place in the world's economy and its lingering environmental dangers.

Krull, Kathleen. *Marie Curie.* New York: Viking, 2007. A modern treatment of the life and work of Marie Curie.

McClafferty, Carla Killough. *Something Out of Nothing: Marie Curie and Radium.* New York: Farrar Strauss Giroux, 2006. A highly readable account of the story of Marie Curie and the discovery of radium.

Mullner, Ross. *Deadly Glow: The Radium Dial Worker Tragedy.* Washington, D.C.: American Public Health Association, 2003. An account of the radium dial tragedy and how it formed the impetus for further use of radium.

Pflaum, Rosalynd. *Grand Obsession: Marie Curie and Her World.* New York: Doubleday, 1989. A modern treatment of the life and work of Marie Curie.

Segrè, Emilio. *Enrico Fermi: Physicist.* Chicago: University of Chicago Press, 1970. The definitive biography of the life and work of Enrico Fermi, written by his good friend and compatriot, Nobel laureate Emilio Segrè.

Internet Resources
Argonne National Laboratory. "Cesium: Human Health Fact Sheet." Available online. URL: www.ead.anl.gov/pub/doc/radium.pdf. Accessed on June 23, 2009. Contains details on cesium sources, uses, and human health risks.

General Resources
The following sources discuss general information on the periodic table of the elements.

Books and Articles

Ball, Philip. *The Elements: A Very Short Introduction.* Oxford: Oxford University Press, 2002. This book contains useful information about the elements in general.

Chemical and Engineering News 86, no. 27 (2 July 2008). A production index is published annually showing the quantities of various chemicals that are manufactured in the United States and other countries.

Considine, Douglas M., ed. *Van Nostrand's Encyclopedia of Chemistry,* 5th ed. New York: John Wiley and Sons, 2005. In addition to its coverage of traditional topics in chemistry, the encyclopedia has articles on nanotechnology, fuel cell technology, green chemistry, forensic chemistry, materials chemistry, and other areas of chemistry important to science and technology.

Cotton, F. Albert, Geoffrey Wilkinson, and Paul L. Gaus. *Basic Inorganic Chemistry,* 3rd ed. New York: Wiley and Sons, 1995. Written for a beginning course in inorganic chemistry, this book presents information about individual elements.

Cox, P. A. *The Elements on Earth: Inorganic Chemistry in the Environment.* Oxford: Oxford University Press, 1995. There are two parts to this book. The first part describes Earth and its geology and how elements and compounds are found in the environment. Also, it describes how elements are extracted from the environment. The second part describes the sources and properties of the individual elements.

Daintith, John, ed. *The Facts On File Dictionary of Chemistry,* 4th ed. New York: Facts On File, 2005. Definitions of many of the technical terms used by chemists.

Downs, A. J., ed. *Chemistry of Aluminium, Gallium, Indium and Thallium.* New York: Springer, 1993. A detailed, wide-ranging, authoritative, and up-to-date review of the chemistry of aluminum, gallium, indium, and thallium. Coverage is of the chemistry and commercial aspects of the elements themselves; emphasis is on the design and synthesis of materials and on their properties and applications.

Emsley, John. *Nature's Building Blocks: An A–Z Guide to the Elements.* Oxford: Oxford University Press, 2001. Proceeding through the periodic table in alphabetical order of the elements, Emsley describes each element's important properties, biological and medical roles, and importance in history and the economy.

———. *The Elements.* Oxford: Oxford University Press, 1989. In this book, Emsley provides a quick reference guide to the chemical, physical, nuclear, and electron shell properties of each of the elements.

Gray, Harry B., John D. Simon, and William C. Trogler. *Braving the Elements.* Sausalito, Calif.: University Science Books, 1995. This book is an introduction to the basic principles of chemistry, with elementary explanations of radioactive decay, chemical bonding, oxidation-reduction reactions, and acid-base chemistry. Practical applications of specific chemical compounds and classes of compounds are presented.

Greenberg, Arthur. *Chemistry: Decade by Decade.* New York: Facts On File, 2007. An excellent book that highlights by decade the important events that occurred in chemistry during the 20th century.

Greenwood, N. N., and Earnshaw, A. *Chemistry of the Elements.* Oxford, U.K.: Pergamon Press, 1984. This book is a comprehensive treatment of the chemistry of the elements.

Hall, Nina, ed. *The Age of the Molecule.* London: Royal Society of Chemistry, 1999. This book is an excellent introduction to contemporary applications of chemistry, including the topics of catalysis, electrochemistry, and the synthesis of new materials.

———. *The New Chemistry.* Cambridge, U.K.: Cambridge University Press, 2000. This book contains chapters devoted to the properties of metals and electrochemical energy conversion.

Hampel, Clifford A., ed. *The Encyclopedia of the Chemical Elements.* New York: Reinhold Book Corporation, 1968. In addition to articles about individual elements, this book also has articles about general topics in chemistry. Numerous authors contributed to this book, all of whom were experts in their respective fields.

Heiserman, David L. *Exploring Chemical Elements and Their Compounds.* Blue Ridge Summit, Pa.: Tab Books, 1992. This book is described by its author as "a guided tour of the periodic table for ages 12 and up," and is written at a level that is very readable for precollege students.

Henderson, William. *Main Group Chemistry.* Cambridge, U.K.: The Royal Society of Chemistry, 2002. This book is a summary of inorganic chemistry in which the elements are grouped by families.

Jolly, William L. *The Chemistry of the Non-Metals.* Englewood Cliffs, N.J.: Prentice-Hall, 1966. This book is an introduction to the chemistry of the nonmetals, including the elements covered in this book.

King, R. Bruce. *Inorganic Chemistry of Main Group Elements.* New York: Wiley-VCH, 1995. This book describes the chemistry of the elements in the s and p blocks.

Krebs, Robert E. *The History and Use of Our Earth's Chemical Elements: A Reference Guide,* 2nd ed. Westport, Conn.: Greenwood Press, 2006. Following brief introductions to the history of chemistry and atomic structure, Krebs proceeds to discuss the chemical and physical properties of the elements group (column) by group. In addition, he describes the history of each element and current uses.

Lide, David R., ed. *CRC Handbook of Chemistry and Physics,* 89th ed. Boca Raton, Fla.: CRC Press, 2008. The *CRC Handbook* has been the most authoritative, up-to-date source of scientific data for almost nine decades.

Mendeleev, Dmitri Ivanovich. *Mendeleev on the Periodic Law: Selected Writings, 1869–1905.* Mineola, N.Y.: Dover, 2005. This English translation of 13 of Mendeleev's historic articles is the first easily accessible source of his major writings.

Minkle, J. R. "Element 118 Discovered Again—for the First Time." *Scientific American,* 17 October 2006. This article describes how scientists in California and Russia fabricated element 118.

Newton, David E. *Chemical Elements: From Carbon to Krypton.* Farmington Hills, Mich.: U•X•L, the Gale Group, 1999. Three-volume set. Despite the title, these books cover the history, physical and chemical properties, natural occurrence and production, uses, and health effects of all of the elements.

Norman, Nicolas C. *Periodicity and the p-Block Elements.* Oxford: Oxford University Press, 1994. This book describes group properties of post-transition metals, metalloids, and nonmetals.

Parker, Sybil P., ed. *McGraw-Hill Encyclopedia of Chemistry,* 2nd ed. New York: McGraw-Hill, 1993. This book presents a comprehensive treatment of the chemical elements and related topics in chemistry, including expert-authored coverage of analytical chemistry, biochemistry, inorganic chemistry, physical chemistry, and polymer chemistry.

Rouvray, Dennis H., and R. Bruce King, ed. *The Periodic Table: Into the 21st Century.* Baldock, Hertfordshire, U.K.: Research Studies Press Ltd., 2004. A presentation of what is happening currently in the world of chemistry.

Stwertka, Albert. *A Guide to the Elements,* 2nd ed. New York: Oxford University Press, 2002. This book explains some of the basic concepts of chemistry and traces the history and development of the periodic table of the elements in clear, nontechnical language.

Van Nostrand's Encyclopedia of Chemistry, 5th ed. Edited by Glenn D. Considine. Hoboken, N.J.: Wiley and Sons, 2005.

Winter, Mark J., and John E. Andrew. *Foundations of Inorganic Chemistry.* Oxford: Oxford University Press, 2000. This book presents an elementary introduction to atomic structure, the periodic table, chemical bonding, oxidation and reduction, and the chemistry of the elements in the s, p, and d blocks; in addition, there is a separate chapter devoted just to the chemical and physical properties of hydrogen.

Internet Resources

American Chemical Society. Available online. URL: www.chemistry. org. Accessed on December 19, 2008. Many educational resources are available online.

Center for Science and Engineering Education, Lawrence Berkeley Laboratory, Berkeley, California. Available online. URL: www.lbl.

gov/Education. Accessed on October 20, 2009. Contains educational resources in biology, chemistry, physics, and astronomy.

Chemical Education Digital Library. Available online. URL: www. chemeddl.org/index.html. Accessed on December 19, 2008. Digital content intended for chemical science education. Chemical Elements. Available online. URL: www.chemistryexplained.com/ elements. Accessed on December 19, 2008. Information about each of the chemical elements.

Chemical Elements.com. Available online. URL: www.chemical elements.com. Accessed on December 19, 2008. A private site that originated with a school science fair project.

Chemicool, created by David D. Hsu. of the Massachusetts Institute of Technology. Available online. URL: www.chemicool.com. Accessed on December 19, 2008. Information about the periodic table and the chemical elements.

Department of Chemistry, University of Nottingham, United Kingdom. Available online. URL: www.periodicvideos.com. Accessed on December 19, 2008. Short videos on all of the elements can be viewed. The videos can also be accessed through YouTube®.

Helmenstine, Anne Marie, Ph.D. "Chemistry." Available online. URL: chemistry.about.com/od/chemistryfaqs/f/element.htm. Accessed on December 19, 2008. Information via the About.com Web site about the periodic table, the elements, and chemistry in general.

Journal of Chemical Education, Division of Chemical Education, American Chemical Society. Available online. URL: jchemed.chem. wisc.edu/HS/index.html. Accessed on December 19, 2008. The Web site for the premier online journal in chemical education.

Lawrence Berkeley Laboratory. Available online. URL: http:// isswprod.lbl.gov/Seaborg/. Accessed on June 23, 2009. A Web site devoted to information about the life and contributions of Glenn T. Seaborg.

Lenntech Water Treatment & Air Purification. Available online. URL: www.lenntech.com/Periodic-chart.htm. Accessed on December

19, 2008. Contains an interactive, printable version of the periodic table.

Los Alamos National Laboratory, Chemistry Division, Los Alamos, New Mexico. Available online. URL: periodic.lanl.gov/default.htm. Accessed on December 19, 2008. A resource on the periodic table for elementary, middle school, and high school students.

Mineral Information Institute. Available online. URL: www.mii.org. Accessed on December 19, 2008. A large amount of information for teachers and students about rocks and minerals and the mining industry.

National Nuclear Data Center, Brookhaven National Laboratory, Upton, New York. Available online. URL: http://www.nndc.bnl.gov/content/HistoryOfElements.html. Accessed on December 19, 2008. A worldwide resource for nuclear data.

The Periodic Table of Comic Books, Department of Chemistry, University of Kentucky. Available online. URL: www.uky.edu/Projects/Chemcomics. Accessed on December 19, 2008. A fun, interactive version of the periodic table.

The Royal Society of Chemistry. Available online. URL: http://www.rsc.org/chemsoc/. Accessed on January 17, 2009. This site contains information about many aspects of the periodic table of the elements.

Schmidel, Dyann K., and Wanda G. Wojcik. "Web Weavers." Available online. URL: quizhub.com/quiz/f-elements.cfm. Accessed on August 13, 2009. A K–12 interactive learning center that features educational quiz games for English language arts, mathematics, geography, history, earth science, biology, chemistry, and physics.

United States Geological Survey. Available online. URL: minerals.usgs.gov. Accessed on December 19, 2008. The official Web site of the Mineral Resources Program.

University of Nottingham. "The Periodic Table of Videos." Available online. URL: www.periodicvideos.com/. Accessed on August 13, 2009. Short videos on all of the elements can be viewed here.

University of Sheffield. "Web Elements." Available online. URL: www. webelements.com/index.html. Accessed on December 19, 2008. A vast amount of information about the chemical elements.

Wolfram Science. Available online. URL: demonstrations.wolfram. com/PropertiesOfChemicalElements. Accessed on on December 19, 2008. Information about the chemical elements from the Wolfram Demonstrations Project.

Periodicals

Discover
Published by Buena Vista Magazines
114 Fifth Avenue
New York, NY 10011
Telephone: (212) 633-4400
www.discover.com
A popular monthly magazine containing easy to understand articles on a variety of scientific topics.

Nature
The Macmillan Building
4 Crinan Street
London N1 9XW
Telephone: +44 (0)20 7833 4000
www.nature.com/nature
A prestigious primary source of scientific literature.

Science
Published by the American Association for the Advancement of Science
1200 New York Avenue, NW
Washington, DC 20005
Tel: (202) 326-6417
www.sciencemag.org
One of the most highly regarded primary sources for scientific literature.

Scientific American
415 Madison Avenue

New York, NY 10017

Telephone: (212) 754-0550

www.sciam.com

A popular monthly magazine that publishes articles on a broad range
of subjects and current issues in science and technology.

Index

Note: *Italic* page numbers refer to illustrations.